TEACHING ABOUT PLACE

TEACHING ABOUT PLACE

LEARNING FROM THE LAND

EDITED BY

LAIRD CHRISTENSEN

AND

HAL CRIMMEL

UNIVERSITY OF NEVADA PRESS

RENO & LAS VEGAS

University of Nevada Press, Reno, Nevada 89557 USA

Copyright © 2008 by University of Nevada Press

All rights reserved

Manufactured in the United States of America

Design by Kathleen Szawiola

Library of Congress Cataloging-in-Publication Data

Teaching about place : learning from the land /

edited by Laird Christensen and Hal Crimmel.

    p.   cm.

 Includes bibliographical references and index.

ISBN 978-0-87417-732-9 (pbk. : alk. paper)

1. English language—Rhetoric—Study and teaching.

2. Report writing—Study and teaching (Higher)

3. Literature—Study and teaching (Higher)

I. Christensen, Laird, 1960–   II. Crimmel, Hal, 1966–

PE1404.T2725   2008

 808'.042—dc22   2007040985

The paper used in this book is a recycled stock made
from 100 percent post-consumer waste materials and
meets the requirements of American National Standard
for Information Sciences—Permanence of Paper for
Printed Library Materials, ANSI/NISO Z39.48-1992
(R2002). Binding materials were selected for strength
and durability.

FIRST PRINTING

17 16 15 14 13 12 11 10 09 08

5 4 3 2 1

CONTENTS

INTRODUCTION
Why Teach About Place?

LAIRD CHRISTENSEN AND HAL CRIMMEL

At the front of nearly any road atlas of the United States is a picture of the nation snared in a colorful net: gray strands run between the states, red strands trace the older roads, and thick blue strands carry shields along the postwar interstates. There are broken lines of time zones, too, and even some faint blue rivers. Many people could point on the map to several places they have called home. Most could name other locations where family members live, and one or two places where they dream of someday living. For some, all these points on the map might be clustered within a fifty-mile radius.

For others, including the editors of this collection, these points span thousands of miles. We each have called more than a dozen places home, and our families are scattered across the nation and overseas. Neither of us inhabits our native places: The one raised north of New York's Adirondacks now lives along Utah's Wasatch Front. The one from the Columbia River's southern shore now lives in Vermont, near the head of Lake Champlain.

This collection is born of that displacement, a response to our decades spent chasing education, adventure, or employment. The tension be-tween the appeal of the road and our desire for roots has defined our lives, personally and professionally, and so we find that our complicated relationship with place informs our teaching and our writing. We are in good company, however, as a similar fascination with place fills the pages of writers such as Scott Russell Sanders, N. Scott Momaday, Janisse Ray, David Orr, Gary Snyder, Kathleen Norris, John Tallmadge, Joy Harjo, Wendell Berry, and John Daniel.

Of course, for most of human history there was no need to question what it meant to belong to a particular place. Survival depended absolutely on people's understanding of the specific possibilities and challenges of the places they called home. So it is not surprising that inhabitory peoples, as Leslie Marmon Silko and Keith Basso have shown, tend to dwell in places that are alive with stories. It is only because we have been "freed" from that most fundamental context by our fossil-fueled economic networks that we find it necessary to remind ourselves that there is something worth learning about the physical places we call home.

Lately, this desire to understand what it means to be a part of a particular place has begun to inspire some of the most innovative teaching in American colleges and universities. Accounts of such learning experiences, however, are rare. We hope that the stories included in this collection might inspire other teachers to discover what may be learned through close investigation of their own places. Though instructors will find many helpful suggestions in this volume, these essays are not intended as templates for particular courses. Rather, our hope is that the narrative approach used here will reveal broader lessons about the possibilities and limitations that come with teaching about place.

Perhaps this collection may also contribute to the larger mission of expanding the way college instructors think about teaching writing and literature. Most of these accounts approach their subjects in experiential and interdisciplinary ways, and the fact that many feature field components reflects a general sense that the best way to learn about a place is on location. Similarly, many of the essays illustrate the value of collaborating with faculty from other disciplines, as place-based studies are especially resistant to artificial (and often arbitrary) academic boundaries.

We hope the collection will assist instructors in the humanities, and perhaps in the social and life sciences as well, in considering how teaching about place is especially relevant in the increasingly mediated educational environment of the twenty-first century. As one anonymous reviewer of this collection observed, "Those of us in higher education could be doing more than we traditionally do to encourage students not to take places—with their unique ecological and cultural attributes—for granted, despite the overwhelming tendency of mass American consumerism and pop culture to obscure or obliterate local differences, and despite the strong tendency of American higher education to neglect the subject of place."

In our attempt to accomplish these goals, we have included essays that model successful courses from across the United States, realizing that any attempt to cover all regions, landscapes, or ecosystems would fall short. The coverage, then, is meant to be suggestive rather that comprehensive. That said, we recognize that the collection's geographical distribution emphasizes the West, partly because place-based scholarship found an early home at western colleges and universities. But the collection's westward tilt may also reflect how the preponderance of public lands in the West demonstrates conflicts about government control, economic opportunities, and private recreation.

Whether focused on western deserts or eastern forests, this collection assumes a special urgency in light of the transformation of place that is occurring most everywhere across our nation. We have all witnessed the loss of special places before the sprawl of human development, and we cannot help but be struck by how much, how fast. How should we react to the promise or threat of change to the places we live? How do we make sense of such changes? Teaching about place provides a starting point for understanding and reacting to these transformations.

A thorough discussion of the concepts of place, region, and bioregion deserves a book unto itself, and certainly requires more space than we have in this introduction. Readers interested in broader issues of place would do well to consult John A. Agnew and James S. Duncan's *The Power of Place* (1989), E. C. Relph's *Place and Placelessness* (1976), and a pair of classic studies by Yi-Fu Tuan: *Space and Place: The Perspective of Experience* (1977) and *Topophilia: A Study of Environmental Perception, Attitude, and Values* (1974). Readers with a general interest in the overlap of story and place should consult *A Companion to the Regional Literatures of America* (2003), edited by Charles L. Crow; the introduction to Michael Branch and Daniel J. Phillipon's *The Height of Our Mountains: Nature Writing from Virginia's Blue Ridge Mountains and Shenandoah Valley* (1998); Lawrence Buell's *The Environmental Imagination* (1995); Kent Ryden's *Mapping the Invisible Landscape: Folklore, Writing, and the Sense of Place* (1993); and Gary Snyder's *The Practice of the Wild* (1990).

Our collection is divided into three sections, of which the first, "Teaching in Place," demonstrates something of the breadth of courses that investigate issues of place on location. In "Calamity Brook to Ground Zero," coeditor Laird Christensen describes a semester spent traveling with three other faculty and twenty students from the Hudson River's source in the Adirondacks to its mouth in Manhattan. Focusing on the watershed as a single system, Christensen's class observes how the battle over dredging PCBs from the channel shapes the way residents understand their relationship with the river. Next, in "Learning Nature Through the Senses," Susan Zwinger and Ann Zwinger explore how careful attention to an expanded range of perceptions helps connect one to place, whether in the dry air of the Grand Canyon or the temperate rain forests of the Pacific Northwest.

In "Uplift and Erosion: Together Along the San Gabriel Front," Bradley John Monsma describes coming to know the Los Angeles Basin through a rain-soaked journey with students to the site of a disastrous mudslide. Understanding catastrophe and change, notes Monsma, helps students to see place as a continual process of creation and loss. Concluding the first section is a contribution by John Elder, who has done as much as anyone to explore the possibilities of teaching about place. Although raised in California, Elder has spent decades discovering his place in the Green Mountains of Vermont. His essay, "A Teacher on the Long Trail," describes his evolution from initial excursions with students in his nature-writing courses to an approach that features increasing "emphasis on civic and practical matters."

Part Two, "Making Connections," demonstrates how teachers have encouraged their students to work across conventional boundaries between communities and disciplines. From the western edge of the Great Basin, Cheryll Glotfelty describes how she transformed a core course at the University of Nevada by incorporating the concept of place into the curriculum. In "Thinking About Women in Place," she reflects on how the intersection of feminist and bioregionalist notions of "home" helped students challenge traditional notions of gender and domestic space. Of course, attaining a nuanced understanding of place requires recognizing its multiple layers, as SueEllen Campbell reveals in "The Complexities of Places." After illustrating how rich, yet incomplete, our casual interpretations of landscape are, Campbell shares a perceptual exercise designed to foster appreciation for the many complex and interdisciplinary possibilities for making sense of place.

We return to the Great Basin, to Utah's Salt Lake Valley, for the next essay, Jeffrey Mathes McCarthy's "A Place at the Table: Writing for Environmental Studies." McCarthy offers his students a remarkably tangible awareness of the region they inhabit by focusing a freshman composition class on a bioregional food-gathering project. In "Meet the Creek," biologist Ellen Goldey and nature writer John Lane lead us to the piedmont of South Carolina, where college students help local schoolchildren appreciate the ecological diversity of Lawson's Fork. Working with a grant from the National Science Foundation, Goldey and Lane demonstrate how instructors can make use of local places to bridge the gap between science and humanities. The concluding essay in this section, "Beneath the Surface: Natural Landscapes, Cultural Meanings, and Teaching About Place," takes us to the Rachel Carson National Wildlife Refuge in coastal Maine. There, Kent C. Ryden helps his students learn to unpack an "authored landscape" by exposing hidden layers of "imaginative, social, cultural, and historical significance."

Part Three, "Meeting the Challenges," offers perspective on the difficulties encountered when teaching about place. John Price begins with his essay, "Idiot Out Wandering Around: A Few Words About Teaching Place in the Heartland." Grasslands too often have been viewed only as places to be broken by the plow or crossed as quickly as possible, notes Price. How, then, might we teach place in a diminished ecosystem, devoid of wilderness and sublimity? Terrell Dixon and Lisa Slappey also work to answer that question in their essay, "The Bayou and the Ship Channel: Finding Place and Building Community in Houston, Texas." Teaching in urban places requires overcoming the notion that cities have "built-in placelessness," note the authors, showing how even a flat, heavily industrialized landscape like Houston offers multiple possibilities for building a sense of place. The theme of recovery and renewal continues in Rochelle Johnson's "Rediscovering Indian Creek: Imagining Community on the Snake River Plain." The essay describes working with students to tell the story of a long-neglected creek in the small city of Caldwell, on the Snake River plain of southern Idaho. Students must learn to see past a starkly degraded landscape before they can begin to find the rewards of "fostering the ecological, emotional, and economic health of a place."

In "Gifts and Misgivings in Place," Paul Lindholdt writes from eastern Washington, on the edge of forested mountains, rolling hills of fertile loess, and shrub-steppe desert. The essay asks us to consider what it means to teach place in the rural West, where individual freedom is exalted and the word "environmentalist" is often used as an epithet. Greg Gordon's essay, "Weaving the Wildness: Exploring the Paradox of Teaching About Wilderness as Place," draws on his years of teaching month-long field programs to help readers see behind facile notions of wild places. Set in the Greater Yellowstone ecosystem, Gordon's essay shows how students immersed in the study of "rangeland conditions, ungulate migration patterns, and wolf predation" come to see the inherent paradox in managing "wild" animals and places. Finally, coeditor Hal Crimmel offers "Teaching About Place in an Era of Geographical Detachment."

Set in Dinosaur National Monument's high desert, the essay considers how regionally focused teaching can help students realize that connecting with place has fleeting as well as lasting influences.

One obvious response to the displacement of contemporary American society is to embrace membership in local communities, which may include expanding the definition of "community" to include the non-human, as Aldo Leopold urged more than fifty years ago in *A Sand County Almanac and Sketches Here and There* (1949). As a biologist, Leopold understood that no organism—not even the cleverest two-legged kind—can possibly exist apart from our fundamental biotic communities, in which organisms and processes cycle the sunlight and water and minerals that sustain us every day of our lives. His land ethic was an attempt to change "the role of *Homo sapiens* from conqueror of the land community to plain member and citizen of it. It implies respect for his fellow members, and also respect for the community as such" (204).

Leopold chose his words carefully, recognizing that citizens have rights but also a responsibility to try and understand how our own shadows fall across a hundred tiny worlds. We may begin to make different choices as the notion of belonging to place acquires both a local significance and a broader one. Does finding satisfaction in the purity of a local creek become diminished when one knows that one's joy is the result of a displaced impact, as industrial pollutants are now simply draining into a stream in China, for example?

Such is the process of bioregional education, which has a rich history in other cultures, perhaps most famously illustrated by the Haudenosaunee tradition of anticipating consequences seven generations into the future. But bioregional education remains a fairly new addition to our formal, Western curriculum. In 1949, Leopold feared that we were not even headed in the right direction: "The most serious obstacle impeding the evolution of a land ethic," he observed, "is the fact that our educational and economic system is headed away from, rather than toward, an intense consciousness of land" (261).

A glimpse into the globalized future might only have heightened Leopold's fears about our economic path, though perhaps he would have found some hope in the grassroots phenomenon of community supported agriculture (CSA), or the growing interest in local and organic produce. Certainly he would have welcomed the recent infusion of environmental education into the national curricula, from elementary to graduate schools. We hope that the essays collected here may serve as further evidence that at least a portion of our educational system is slowly finding its way toward that "intense consciousness of land," and that they may help us all, students and teachers, begin to rediscover the value of living and learning in place.

PART I ✑ TEACHING IN PLACE

CHAPTER 1

Calamity Brook to Ground Zero

LAIRD CHRISTENSEN

The afternoon is balmy for November, and the students aboard River-
keeper's patrol boat have shed their jackets as they ride the current of
Rondout Creek toward the Hudson. The creek has left an impression
on them. From the barge-cluttered banks to the hillside ranks of con-
dominiums, this battered creek looks nothing like the watershed they
have come to know in the Hudson's upper reaches. Here, they glide past
gravel mines and industrial boatyards. By the time they drift past heaps
of flattened automobiles lining the western shore—where antifreeze,
gas, and transmission fluids drain directly into the creek—they look
slightly dazed.[1]

This creek would not seem out of place in most industrial landscapes,
but it alarms these students. Since the end of August they have been
working their way south from the Adirondack headwaters of the Hudson, learning to see the entire drainage as a single organic system. It is
the only subject they will study this semester. The students have watched
the river lose its innocence as they have paddled, sailed, and driven
downriver toward the twenty-first century. They have learned about

the lives of local inhabitants—both human and not—and pictured the Hudson through the eyes of Mahican traders and Romantic painters. Here, at Rondout Creek, they see it as a tool, a highway, a drain.

Try making that point in a traditional classroom.

Courses like this, in which students spend a block of fifteen credits studying a single place through overlapping disciplines, are central to the mission of Green Mountain College. Designed to engage students through local relevance or current controversy, these upper-level courses embody the school's environmental approach to liberal arts education. On their weekly trips into the field, as well as in the classroom, students experience knowledge that is integrated and applied—a way of knowing that will help them think critically about the lives and homes awaiting them beyond commencement.

As our college's neighboring watershed to the south and west, the Hudson was an obvious choice for our 2001 block course. It is, after all, the nation's largest Superfund site. Even as the Environmental Protection Agency was weighing options for cleaning up the river, communities up and down the Hudson splintered into shouting matches. The twenty students enrolled in the course had only to cross the Poultney River to see the argument rage in signs along New York roadsides. Some residents, along with downstream activists, insisted that the General Electric Corporation must pay for dredging the riverbed of spilled carcinogens—the notorious polychlorinated biphenyls, or PCBs. Others argued that dredging would simply stir up the chemicals, making matters worse. The scorched remains of a large anti-dredging banner in Fort Edward, New York, registered the heat of local emotions.

From the moment my colleagues and I began planning this course, however, we found ourselves wading into the unknown. To begin with, this was not our neighborhood. Not even close. Of the four of us teaching the course, none was from any farther east than Iowa. We had learned our trades at doctoral programs in Arizona, California, Oregon, and Colorado—three of us quite recently. We had been drawn to Green Mountain College by the dream of place-based teaching, but we were long on theory, short on practice.

Beyond that, I had my own doubts—common, I think, to those trained in writing and literature—about what I could contribute to such a course. It made sense having a biologist on board, and Dr. Meriel Brooks had already taught a similar course on the Champlain Basin. And certainly students could use a social historian like Dr. Patricia Moore to help them understand how the river's human communities had evolved since European colonization. Dr. Jon Jensen, from environmental studies, would help untangle the web of policies and agencies that regulate the river's health. But was the region's literature really an essential component of the course?

Wouldn't a geologist have made more sense?

Our studies began at the headwaters, high in the blue Adirondacks. Under a cloudless late-summer sky, we pitched camp near the village of Newcomb and set out to learn the neighborhood. Dick Sage of the Adirondack Ecological Center helped us to make sense of the region's ecology, leading us from one stage of forest succession to the next. Later, we traveled upstream to where the Hudson, running ankle-deep through spruce and balsam fir, is still known as Calamity Brook. Some of our students back-tracked the current all the way up to Lake Tear of the Clouds, spending an extra night on the trail.

Wading the current and camping along the banks, we grew comfortable with the river and gradually invested in its fate. These quiet days set the stage for a dramatic contrast. As student Megan Fries wrote in her journal, "We were to see how beautiful the river is in this spot so that down the road we could understand what has been done to it by the years of human interaction."

Wherever I teach literature, I begin with the stories that grew out of the local landscape. In the case of the Adirondacks, that meant *Tales of the Iroquois,* by Tehanetorens—one of many recommendations from Joseph Bruchac, who kindly guided me toward a basic understanding of the Hudson's native cultures, and even came to Vermont to share his stories with our class.[2] Not only can such stories bring the landscape to life, but introducing students to the economy and social structures of indigenous populations helps them to imagine what a more sustainable form of inhabitation looks like. They also read John Burroughs's essay "The Adirondacks," Anne LaBastille's *Woodswoman,* and Bill McKibben's description of resurgent wildness from *Hope, Human and Wild.* LaBastille's book in particular offered a very personal account of learning to be at home in this remarkable landscape.

While literature added an emotional layer to our studies, however, I found that our common goals for the course were less well served by sustained literary criticism than by finding connections between the reading and other disciplines, as well as the students' own experiences. At times, each of the instructors was frustrated by the lack of disciplinary depth we were able to achieve, though we ended up marveling at the breadth of synthesis that emerged. Along the way, I also discovered that other aspects of my training were more urgently needed than literary analysis.

In Newcomb, we met with George Canon, the florid, stocky town manager who had worked thirty years in the nearby Tahawus Mine. His early-morning account of local history unexpectedly picked up steam as he veered into a tirade against state ownership of land in Adirondack Park. "It's no different from communism," he declared. At this, the students began to pay a different kind of attention. Curious and alert, they laid aside their pens. As the state buys up the land, Canon continued, unemployment climbs, and "it's asinine to just let the trees rot in the forest." Yes, there's more tourism, he admitted, but "hikers and tree huggers don't tend to spend much money." Even as the students chuckled at their own empty pockets, I was scribbling a note to myself: "Teach rhetorical analysis!"

After Canon's talk we explored the old titanium mine, scampering up the mountain of tailings where a young Hudson is diverted through a narrow canal. It was difficult to accept Canon's assurance, still fresh in our ears, that mining had not degraded the site. The students could not so easily dismiss his claim, however, that current policies reward downstate vacationers at the expense of the local economy. We had spent so much time discussing the virtues of the wild Adirondacks that it was bracing—even enlightening—to hear another point of view proclaimed with such conviction.

Later, as we canoed through Warren County, the landscape seemed to age centuries in only a few miles. We began beneath the sweeping white cedars, where students gathered water samples for Meriel Brooks and watched a bald eagle rise from the river, a trout in its talons. Soon enough, however, we were gliding through the mud from crumbling banks at the edges of developed land, then paddling through the fumes of a grumbling generator. By the time we skirted the manicured edges of a golf course, unnaturally green, we knew this was a different Hudson.

⟢

We were back on campus when the news broke. I overheard mention, while paying my breakfast bill at the diner, of a plane crash, but I had other things on my mind. The conversation at the convenience store was more urgent, and I wondered why people were so worked up about an accident. "Did you hear?" one colleague asked, then another, as I arrived at school. It was the strain in their voices that sent me first to the Internet, then to find a television.

In the student lounge I joined the dozen or so who were crowded, hushed and anxious, around the large screen. We watched the black smoke billow from the first tower. Then we saw the jet—streaking across the screen with unfathomable deliberation—slam into the second. We watched it happen again. Then again. What came next, of course, eclipsed the horror of those burning buildings. The shudder, the buckling, and then the oddly geometric collapse, tiers of windows flicking past like the shuffled corners of cards.

After watching the second tower fall, and then watching the entire scene again, and again, and again—as if waiting for sense to be made—I turned away at last, finally numb.

How does one go on with daily tasks in the wake of such tragedy?

By embracing those very tasks, perhaps, which give us *something* to be doing. Not to have held class in those first few days would have denied us all a place, a community, left over from the world we used to know. It mattered less that we were productive, or on task, than that we simply met.

In the weeks that followed, it seemed that the staff and faculty were having a harder time coping with the attacks of September 11 than the students were. Maybe it was just that we've been around long enough to have more comfort invested in a suddenly outdated notion of domestic security. Regardless, it was hard to think about the course in quite the same way as before. After all, each week's field trip would lead us closer to the mouth of the Hudson, and to what we were now calling Ground Zero. One other problem: how important would students find the threat of PCBs compared to the sudden deaths of thousands downstream?

There was no sign of fall in the last morning of summer, as our vans eased through the frayed town of Fort Edward, New York, and banked down to the local marina. Students stretched and yawned as they piled out, drifting toward the riverbank. The marina had by now become a familiar classroom. Just upstream, where two GE plants filled transformers into the 1970s, more than a million pounds of PCBs had leaked into the river. From here we could see the riffle where the Allen Mill Dam once stood. Its removal in 1975 released millions of tons of toxic sediment downstream toward Albany, Manhattan, and the Atlantic.

The students gathered in a small gazebo as an oversized pickup rolled down the drive. "That must be Tim Havens," one student supposed, and others reached for notebooks to review their questions for this morning's speaker. A beefy, bespectacled man eased out of the cab in work boots, a taut plaid shirt, and a baseball cap that read, "Stop the Dredging." Tim Havens, who deals in farm equipment for a living, had become president of CEASE (Citizen Environmentalists Against Sludge Encapsulation) in 1980, at the tender age of eighteen. For more than an hour, there in the gazebo, the opposition to dredging had a face and a voice, a family and a home on the river. Students fidgeted as Havens tossed out pat answers to their careful questions. Identifying himself as "a true steward of the land," he assured the students that PCBs are only "a *suspected* human carcinogen." Dredging would simply stir up the chemicals, he insisted. "They're in a watery tomb. They aren't coming back."

Driving south from Fort Edward, we passed the Thompson Island Pool, a willow-shaded stretch of river that has the highest concentration of PCBs on the Hudson—so high that the bodies of fish-eating animals found nearby have been treated as hazardous waste. "How can such a beautiful river be so toxic?" wondered a student as we rattled past. The question hung in the air even as we gathered on the lawn of the historic Schuyler House, a dozen miles downriver, to hear another point of view.

With a nod to David and Goliath, Sierra Club activist Chris Ballantyne described the struggle to compete with GE—which spent, he guesses, $25 million making its case against dredging. The Sierra Club had become involved rather late in the game; its presence was needed, Ballantyne explained, because "there has never been an effective citizens' group advocating the cleanup." He seemed to enjoy swatting down the points raised by Havens an hour before. Then, like a man running for office, he passed around buttons and stickers announcing support for the dredging. Once again the students squirmed. Determined to observe the debate objectively, some took offense that Ballantyne seemed to assume they shared his position. Others welcomed his arguments, glad to find themselves back in a world of simple answers, where good guys force the bad guys to clean up their mess.

Meeting with activists and residents up and down the Hudson brought to life the arguments we had discussed back on campus. When voiced by people directly affected by the outcome, those arguments became inextricable from each tangle of personal circumstances. Even students who leaned toward dredging disliked the thought of downstate activists forcing change on a largely resistant local population. Others were more concerned that area residents, their own health in danger, were being manipulated by corporate propaganda. Because we had by now practiced uncovering the premises of competing claims and recognizing logical fallacies, they were listening more critically than before and responding with precision.

My increased emphasis on rhetorical analysis did not mean, however, that I had given up teaching literature. We were in Mahican country now, and students had read the story of Jodikwado, the river serpent who rescued a young wife from the treachery of her sister-in-law.[3] They had also read T. C. Boyle's novel *World's End,* which was the most useful book I assigned. By telling the interlocking stories of several Hudson River families over four hundred years, Boyle succeeded in bringing dim history to life, supplying human faces for Dutch patroons, indentured servants, and what remained of the native populations. A healthy blend of magical realism, drugs, and sex kept students moving through Boyle's darkly comic prose. It was a wonderful coincidence that he described with such precision and wit some of the very places we were visiting—and even the ship our students knew so well.

Slicing upriver beneath 4,000 feet of canvas, the *Clearwater* slipped like a ghost between the hulks of modern tankers and the low dome of the Indian Point Nuclear Power Plant. This wooden sloop is a replica of an 1859 cargo vessel, designed specifically to fit the quirks of the Hudson. Since it was launched in 1969, the *Clearwater* has provided environmental education to thousands of schoolchildren each year. Over the course of the semester, each of our own students spent a week, in pairs, living and working on board. They helped raise the 3,000-pound mainsail every morning, took water samples, and taught elementary students about the river. On this breezy October afternoon our entire class boarded the *Clearwater* at Haverstraw Bay—the widest part of the Hudson, three miles from bank to bank—and sailed up to where the river narrows and runs deep through the Highlands.

Even as the students took turns crewing on the *Clearwater*, they were recalling the same vessel as portrayed in *World's End*. In fact, the *Arcadia* (as Boyle dubs an identical sloop) is the stage for the novel's climactic scene. One inspiration for the *Arcadia*'s environmental mission is a venerable folksinger named Will Connell, whose role mirrors that played by Pete Seeger on the *Clearwater*. In his journal, student Christian Meny recalls Seeger's visit to the *Clearwater* in the days following the attacks on the World Trade Center: "The wise old man, banjo in his hands, wisdom in his eyes, safety in his presence, compared the cooperation and love that the world needs," writes Meny, "to that that has been given to the Hudson River by those on the sloop *Clearwater*." There's something magical about that moment when the line between

fact and fiction dims and nearly disappears. We find ourselves inhabiting a world made of stories, and simultaneously inhabiting the stories
made of places we know.

We also explored the mid-Hudson ashore, trying to imagine its appeal to Romantic artists. We read essays and poems by Thomas Cole and
William Cullen Bryant as we visited the spots that inspired the Hudson
River painters, as well as Frederic Church's home at Olana. Each student owned a book that included nearly seventy prints associated with
the Hudson River School, and we saw some of the originals—including
works by Cole, Asher Durand, and Albert Bierstadt—at the Marsh-
Billings Rockefeller National Historical Park in Woodstock, Vermont.[4]
At the site of the Old Catskill House, students worked with pastels
to produce their own landscapes. While in the neighborhood, we also
followed the steps of Rip Van Winkle up Kaaterskill Creek and found
the rusty woods as charming—and nearly as charmed—as Washington
Irving had led us to imagine they are.[5]

The weather stayed mild midway through November as we followed the
river toward New York City, crossing at Tappan Zee and merging into
the stiffening traffic of Yonkers. At the Beczak Environmental Education Center we learned about the staff's work with urban youth, some
from Latino and Asian families who cannot decipher the signs warning
against fish consumption. But the real education was seeing, in the fad-
ing daylight, a new face of the Hudson. Fenced off from the public, the
only river access in Yonkers was a tiny beach, littered with bleach bottles
and other scraps of trash. After touring the center we stepped back out
into autumn's sudden evening to find a low, glowing sky clamped over

the river, humming its urban monotone. This may have been the same watershed we'd been studying for months, but we were in a different world now.

The following morning we visited the offices of the Environmental Protection Agency on the twenty-ninth floor of the federal building in Manhattan, five blocks east of where the Hudson chafes between banks of concrete. Doug Tomchuk, who had managed the Hudson River cleanup project since 1989, presented the agency's research in generous detail and took student questions long past our allotted time. We could not have known then that the agency's official decision about how to address the contamination was only three weeks away, but Tomchuk was clearly ready to see this step completed.

After lunch, a wind-whipped ride on the Staten Island Ferry showed us the mouth of the Hudson from New York Harbor, which we struggled to connect with our memories of Calamity Brook. At Battery Park students described to their journals an overwhelming human presence: helicopters chopping through the rare parcels of sky, passersby speaking a dozen different languages, horns from cars and ferries alike, the sounds of heavy equipment at Ground Zero, the remaining skyscrapers looming over us. Here the river seemed less the defining feature of the ecosystem than an afterthought. "The millions of people who walk down the streets of New York City every day," wrote student Kathryn Crane, "most likely aren't even aware of the correlation between their lifestyle and the Hudson River." Indeed, it was hard for us to recall that this park where we sat writing was the tip of an island.

At that time there were no viewing platforms above Ground Zero, so we peeked past sheets of plywood that covered the cyclone fences. Some students found the visit morbid; others were drawn to the site like pilgrims. The most poignant images were the handmade posters that covered the nearby walls: "Have you seen this person?" they asked again and again. Some of them were clearly new. The photos tended to be candid shots, torn from some happier moment and pasted above descriptions of what the victims were wearing that morning, two months before, as they headed out the door. I was stunned to realize that, even this long after the attacks, families were clinging to any hope they could muster.

Neither the exuberant anticipation of Walt Whitman—"Just as you feel when you look on the river and the sky, so I felt"—nor the graceful good nature of E. B. White could lift the pall from our time in Manhattan.[6]

The faces on those posters stayed with me long after we returned from the city. It would be fair to say I was haunted. Stirred by the evidence of so much loss, I was grateful for the peaceful life that my partner and I shared on the shore of Lake St. Catherine. I suppose Vermont has often been a safer place than most, but now I felt that safety every day. I could no longer take it for granted.

On our tiny brick campus, where the Poultney River drains north to Lake Champlain, our study of the Hudson was nearing an end. As students gave formal presentations and prepared to host a public forum on the dredging debate, the EPA released its Record of Decision: GE was ordered to spend half a billion dollars to fund the dredging of contaminated sediment. Some students greeted the decision as a victory, plain and simple. Others worried, with fresh empathy, about the new risks that will come with dredging, separating, and shipping the contaminated sediment through Hudson communities.

We all felt good about the class: the students had gained an unusually broad perspective of the river and its issues, as well as analytical skills that would serve them well wherever they choose to live. They had begun to understand that with the pleasures of knowing a place so intimately come certain responsibilities. They had, in fact, learned the real cost of belonging to community.

On the evening our students hosted the public forum I stayed in bed, laid low by a virus, but I was glad for Jon Jensen's report of their good work:

> Their hour onstage stretched toward two as they fielded questions about PCB chemistry and led the audience through the maze of Superfund law. At the conclusion of the forum, students arranged themselves at the front of the room according to their own opinions about dredging the Hudson. With peers and neighbors looking on, they formed three lines— sitting, standing, and raised onstage—separating those who agreed, disagreed, and remained undecided as to the wisdom of the EPA's decision. After four months, thousands of miles, and countless hours of study, it came down to this: twenty young faces gazing out from the stage, each announcing a personal stand.[7]

I was proud that they displayed such diversity of opinion, even after a semester of studying identical information.

But was this enough? The elephant in the room that nobody mentioned, of course, was the tragedy of September 11. How could it not have shaped our experience of the Hudson, when in fact it had shaped the rest of our lives?

Perhaps we should have been clearer about how the attacks fit into our study of the watershed. I am an instructor who regularly suspends my syllabus to make room for current events relevant to the learning goals, and there is no question that the events of September 11 had a direct impact on the lower Hudson watershed. Perhaps we could have incorporated a section on testing the air quality around Ground Zero, as we had noticed the EPA sensors mounted above our heads, gathering data. Or we could have devoted a final portion of the class to brainstorming appropriate replacements for the twin towers.

We might even have considered possible parallels between the past and present. Are there legitimate comparisons to be made between European incursions up the Hudson, we could have asked, and the contemporary American influence in the Middle East that inspired the attacks? When colonial powers sailed up the river then known as Muhheconnuk four centuries ago, furs and timber were the oil of the day. How precise are the similarities, we might have asked? At what points do they diverge?

We could have taught it that way. But I, for one, didn't yet have the heart.

Perhaps we did enough simply by modeling an experiential approach to making sense of issues that divide a community. We became thoroughly familiar with the contexts, both cultural and scientific, and made sure to hear from people on both sides of the conflict. We analyzed critically all that we heard, and made sure that each argument's premise was clear. Only then, after evaluating all the information, were we ready to reach our personal decisions.

There is something in that approach that suggests a humbler alternative to the kind of decision making that sends airliners into office buildings—or armies into Baghdad. Something that sounds a little like hope, and maybe even like healing.

NOTES

1. Portions of this essay originally appeared as "A Hudson River Immersion," cowritten with Jon Jensen, in *Whole Terrain: Reflective Environmental Practice* 13 (2004/2005): 9–13.

2. I am especially grateful to Joe Bruchac, who made time to meet with me in his home and provided me with maps and handouts that I later used to help students understand who lived where along the Hudson. Among the references to which he directed me, I found two especially helpful: Volume 15 of *The Handbook of North American Indians* (edited by Bruce G. Trigger for the Smithsonian Institution in 1978) and the two volumes published in 1872 by E. M. Ruttenber: *Indian Tribes of Hudson's River to 1700 and Indian Tribes of Hudson's River 1700–1850,* which were reissued by Hope Farm Press in 1992.

3. The story "Jodikwado and the Young Wife" is taken from Joseph Bruchac's collection *Return of the Sun: Native American Tales from the Northeast Woodlands.*

4. The faculty had not intended to require students to purchase such a volume, assuming that the expense would be prohibitive, until we discovered Bert Yaeger's folio-sized collection, *The Hudson River School: American Landscape Artists.*

5. Much of the essential literature of this region, including Irving's stories, is compiled in Arthur G. Adams's collection, *The Hudson River in Literature.*

6. Adams's collection includes two of Whitman's poems, one of them "Mannahatta," from which this excerpt is drawn. Students also caught glimpses of the city through some of E. B. White's contributions to the *New Yorker,* collected and published by Rebecca M. Dale in 1990.

7. Jon Jensen's description of the forum appears on page 13 of "A Hudson River Immersion."

COURSE READINGS

Adams, Arthur G., ed. *The Hudson River in Literature.*

Adams, Richard. *Legends of the Delaware Indians.*

Allan, J. David. *Stream Ecology: The Structure and Function of Running Waters.*

Boyle, Robert H. *The Hudson River: A Natural and Unnatural History.*

Boyle, T. C. *World's End.*

Bruchac, Joseph. *Return of the Sun: Native American Tales from the Northeast Woodlands* excerpt.

LaBastille, Anne. *Woodswoman.*
McKibben, Bill. *Hope, Human and Wild: True Stories of Living Lightly on the Earth,* excerpt.
Outwater, Alice. *Water: A Natural History.*
Stanne, Stephen, et al. *The Hudson: An Illustrated Guide to the Living River.*
Sullivan, Robert. *The Meadowlands.*
Tehanetorens. *Tales of the Iroquois.*
White, E. B. Selected essays.
Yaeger, Bert D. *The Hudson River School: American Landscape Artists.*

WORKS CITED

Adams, Arthur G., ed. *The Hudson River in Literature.* 2nd ed. New York: Fordham University Press, 1980.

Boyle, T. C. *World's End.* New York: Penguin Books, 1987.

Bruchac, Joseph. *Return of the Sun: Native American Tales from the Northeast Woodlands.* Freedom, Calif.: The Crossing Press, 1990.

Burroughs, John. *Deep Woods.* Ed. Richard F. Fleck. Syracuse, N.Y.: Syracuse University Press, 1998.

Christensen, Laird, and Jon Jensen. "A Hudson River Immersion." *Whole Terrain: Reflective Environmental Practice* 13 (2004/2005): 9–13.

La Bastille, Anne. *Woodswoman.* New York: Penguin Books, 1976.

McKibben, Bill. *Hope, Human and Wild: True Stories of Living Lightly on the Earth.* New York: Little, Brown, 1995.

Ruttenber, E. M. *Indian Tribes of Hudson's River: 1700–1850.* 2nd ed. Saugerties, N.Y.: Hope Farm Press, 1999.

———. *Indian Tribes of Hudson's River to 1700.* 2nd ed. Saugerties, N.Y.: Hope Farm Press, 1998.

Tehanetorens (Ray Fadden). *Tales of the Iroquois.* Summertown, Tenn.: Book Publishing Company, 1998.

Trigger, Bruce G. *The Handbook of North American Indians.* Vol. 15. Washington, D.C.: Smithsonian Institution, 1978.

White, E. B. *Writings from the* New Yorker: *1925–1976.* Ed. Rebecca M. Dale. New York: HarperCollins, 1990.

Yaeger, Bert D. *The Hudson River School: American Landscape Artists.* New York: TODTRI Book Publishers, 1999.

CHAPTER 2

Learning Nature Through the Senses

SUSAN ZWINGER AND ANN ZWINGER

College students are desperate for sensation. Through the news, we hear some of what they choose to do in their free time: dance to pulsing rhythms, create art, climb mountain peaks, get drunk, run whitewater, travel to sunny Mexico, try psychedelics, explore each other's bodies, taste intense new cuisines from other cultures, play body-bruising sports, wear exotic fragrances, color hair with strange hues, and drive really fast. I, Susan, remember dressing exotically, eating organic foods, and exploring the *duende* of deep flamenco passion, playing the guitar on hilltops on moonful nights.

College and university students' preoccupation with sensuality is no accident. Our senses are a direct biological means to understanding the world around us and our relationship with other beings. Living in the sensual is a desperate search to know the self by knowing our animal nature. Students, like the rest of us, evolved as explorers, inquisitive pokers, and hunters. Some students, often the more intelligent ones, are the most reckless in their experimentation. And each year we lose a few, usually male, to extreme sensory exploration combined with naiveté, the result of a nature-less upbringing.

Our classrooms not only deny the original cave and forest but work to create the most efficient, rational populace in the world. Indeed, Western culture, including college curricula, attempts to obscure our evolutionary dependence on our immediate senses.

But understanding nature's complexity requires that we relearn how to use not just the five main senses but also the more than fifty-seven others, senses that animals use and that remain available to us as well. The edge of our consciousness marks both the edge of our sense perception and the edge of our knowledge about the places around us.

College curricula treated my brain's corpus callosum like a brick wall. Looking back on my own experience as a college student, I realized that I evolved Janus-like, with two faces and two separate life experiences. On one side, my intellect was greatly stimulated by literature, philosophy, and art history classes. I loved such classes, and my mind thrived. With the other side, I was mountain-climbing, backpacking, running whitewater, and painting nature. This thriving, sensory, active side seldom crossed over to my academic side, except when poetry flashed an accidental fusion.

In 1969, my senior year at Cornell College, many of us read Edward Abbey's just published *Desert Solitaire* and experienced spontaneous combustion, as young people continue to do upon reading Abbey. Ten of us piled in an old station wagon and drove all day and all night to Canyonlands, a newly created national park.

In one of the wondrously synchronistic phenomena of my lifetime, when I was twenty-one I met an unusual young man from Ohio; his name was Gary, and he had just entered Cornell College as a freshman at the age of seventeen, a year younger than most freshmen. His experience with nature up to that time consisted of wildflower paintings and good poems about midwestern hills and forests. He had never been out West. We naturally included him in our group of novice and experienced backpackers.

What a life-changing trip we had. Because the park was brand-new, trails had not yet been properly signed, archaeological treasures lay out in the sun, and water sources were not marked on maps. We backpacked for ten days in total wilderness in the as yet unscathed and untrekked Canyonlands. Five days out in the backcountry, we got lost, ran out of water, patched up blisters, got confused, and scared ourselves silly with stories around a campfire. A sudden storm dropped two feet of snow, promising an end to our water problem. After melting snow for a couple of hours, I hiked up a high ridge alone and came face-to-face with a female mountain lion. We found ancient Anasazi treasures piled near cave sites (we left them in place) and stared down two thousand vertiginous feet to the confluence of the Green and the Colorado. We were miserable, scared, frozen, dehydrated, miffed, confused—and utterly transcendent!

The young Ohioan with us, who had never experienced glorious, wide-open western vistas, was transformed forever. Gary Nabhan claims that this ten-day trip changed his life. In turn, he has since changed the entire world for the better as a leading ethnobotanist, MacArthur fellow, and World Seed Bank gatherer.

Ironically, our swirling, voracious, filthy-clothed, unplanned entrance and exit of my parents' house in Colorado Springs on our way back to Iowa (thoughtless, no doubt) changed Ann Zwinger's life as well. She could sense that we had found Truth out in those distant canyons, and she wanted some.

❧

Since then, Ann Zwinger has written more than twenty-five natural history books, paddled down dangerous whitewater in a Sportyak, rafted the Colorado fifty or more times, and been dubbed the "Thoreau of the Rockies" by Abbey himself. I, have gone to become an artist and curator of fine arts in Santa Fe, and have written five books of natural history and hundreds of articles and poems. Both of us have taught hundreds of workshops and college courses, a few of them (when we are very lucky) together. She and I have been sharing wilderness experiences, with young and old alike, ever since. And while we have found many ideas to share, we are a generation apart and have different brain halves dominant. As one might expect, we have distinct approaches to teaching people a sense of place in the wilds. Still, her writing and teaching have shaped my own, as my poet skills have shaped hers.

We both share the distress over the detachment that students feel from the physical world. During Ann's Colorado College course "Writing the Natural History Essay," which she most recently taught in the fall semester of 2005, she suddenly became aware of this detachment. About her newer students, she wrote:

> I sensed a lack of connection with the natural world I'd not seen before. . . . The generation of parents who now have children in college has been one in which there has been perhaps more emphasis on getting and spending, late and soon, managed childhood, all the pop words we're hearing about us. My epiphany came when I realized that the students I'm teaching now are different, and the difference is that there weren't

as many parents to take a small hand in theirs and watch a dragonfly emerge, giggle at water striders, talk about beavers building their dam across the pond, answer "What is this?" and the endless "Why?"

As an educator, I find that my own approach to teaching natural history in the out-of-doors grew from just such intensely sensory moments, as well as from my hands-on experience as a camp counselor, outdoor educator, wilderness trip leader, instructor for Audubon and National Park institutes, and interpretive ranger in Alaska.

As a professional writer, I know that my own literary peaks result from sensory overloads out in the field of nature. Favorite passages sprang from a night in tidal rapids of British Columbia, getting lost in the Sonoran Desert, summiting Pikes Peak at age fifteen, weeping for a loon caught in fishing line on Whidbey Island, and backpacking solo in the North Cascades for eight days without seeing another soul. Such essential lessons should be shared with students. They must learn that writing about place—in that specific place—will trigger all their senses.

Because I have taught from Maine to Southern California and Alaska to New Mexico, these techniques apply to any particular locale on the planet. However, I am now in love with the Washington, Oregon, and British Columbia forests extraordinaire. Columnar Douglas firs and western red cedars tower two hundred feet over the forest floor. These ancient giants harbor thousands of insect, bird, fungus, and mammal species, a distinct ecosystem at each level of the tree: from bole to canopy. Northwestern old-growth forests produce one of the grandest biomasses per acre in the world, greater than South America's tropical rain forests' 1,163 tons per hectare. I love the Olympic coastal rain forest, where one finds amanitas the size of dinner plates and where fragile voices of frogs and wrens trill unseen in the underbrush. Leaves and needles drip constantly, collective drops, wide dough-balls of sticky rain, exploding the needle carpet when they fall. Even when it has not rained, the forests collect moisture in the form of fog and hold it for the drier summer months. Flying squirrels and other rodents readily eat the false truffles and other fungi of the forest floor, ingesting the spores. Spotted and barred owls, raptors, or larger mammals eat voles and flying squirrels, then scamper or fly the spore to newer forests to inoculate a new forest floor. The mycorrhizae of these symbiotic fungi bring water, minerals, and connection with others to their connected trees. The trees in turn, give them sugar, translating the energy of the sun to the dark, dank mushroom world.

Add to all this the sensations of brushing through head-high ferns and bright yellow skunk cabbage with leaves large enough to wear as a bathing suit, and it becomes evident that the Northwest's ecosystem is interconnected and elegant. I love taking students out into an old-growth forest because the very immensity of it humbles them, entrances them.

When they discover the raindrop smears in my large illustrated journals, students inevitably ask, "Do you actually *sit out there in the cold, dark, and wet* and write?" Yes, emphatically, yes and yes. I stay for hours in place, recording not just the scientific data but the poetry of place. All the senses I can experience are carefully recorded, some for which we have few words, such as smell, recorded in metaphor or synaesthesia. My mother, on the other hand, takes notes very rapidly on large yellow tablets, making plant lists first, then recording detail to type up later. She is trained as an art historian. I am, for better or worse, a poet.

In a classroom, the cerebral processing alone cannot cut to the quick, trigger the emotions, and transform a student's values and life as nature can. Thus, my first step is to teach my students to pay detailed attention through their senses. Senses hold a great power to trigger ancient parts of the brain. All good writers use these ancient brain parts to change fleeting memories into transformative experiences, then into the wisdom of survival. Nature's poignancy has its basis in brain structure. Wild scenes trigger strong emotions. Emotions light up the amygdala. The amygdala, in turn, excites the hippocampus. The hippocampus retains facts, sequences of events in short-term memory, with the potential change to long-term memory. Add emotional response, and short-term memory becomes long-term memories, profound memories. Profound memories lead to paradigm shifts and behavioral changes for the rest of one's life. Thus, the more mountaintop epiphanies, lowland raptures, immersions in river rapids, restless overnights in clammy sleeping bags, tear-filled departures, the more profound the learning.

To achieve this with students I first assign my favorite writing-in-location exercise, which requires them to go out into nature after dark, alone, and record what they experience through as many senses as possible. These include kinesthetics, balance, compass directional sense, Jacobson's organ (which senses pheromones), electromagnetic fields, vibration in earth, air pressure, and vertigo. In addition, they perceive the world through peripheral vision, the olfactory landscape, perioperception, camouflage-perception, subsonic vibration, textures, and echolocation. They become sensitive to air temperature, wind speed and direction, height or depth, pain, air currents, natural rhythms, water currents, gravity, light levels, harmonics, and trace chemicals wafting about. They will touch tongue to bark, bite a needle, and savor a wild leaf.

Sensual exercises work very well in whatever ecosystem serves as an outdoor classroom, including stark desert environments. But travel with me down a deep tunnel of rain forest canopy, through Modigliani roots, to the Pacific Ocean. "Jungle" birds call unseen from the thick, towering canopy. The forest floor is so dark, yet it flickers with patches of light where the sun filters down through the canopy with the intensity of laser beams. The atmosphere is thick with red cedar odor. Ferns are greener than green, as if lit by a miniature lamp under each frond. Cedar planks over the standing black water are slick with algae and rain. A sustained ring, as of a finger on a wineglass, sounds from the underbrush, then another and another, on different pitches. You glimpse the sources of the voice—a dark orange thrush with a black chevron on his chest and intricate wing camouflage: a varied thrush.

Finally, you hear the ocean waves pounding far greater than your small human heart does, and you emerge on a beautiful, rocky beach and jigsaw cove cut off by rocky headlands. You pick your way over the rock shelf, which is worn into hundreds of pools, each filled with colorful creatures: immense sea stars to weirdo crimson dories, to pink, red, and chartreuse crustose algae. Behind you, the great dark rain forest towers, so that if you do not memorize your opening you will never make it back home. You are completely cut off from everyone else at high tide. You could take off your clothes and dance in the highest surf, but it is way too chilly.

Sitting outside at night, with decreased dependence on sight, out in the dark, rain, fog, snow, and a sense of danger will push students' perceived safety edges. Groans of resistance and fear delight me. They are positive signs that folks are being pushed. Danger, or a sense of it, heightens attentiveness, which leads to perceptive, unusual writing.

Some students will hang close by. Some will complain bitterly to themselves for the first twenty minutes, then write something fine in the last ten. Some will walk tentatively out into the trees. Others will stride off through meadows, using their legs to feel the shape of the earth. (Do not do this in bear country.)

Giving the task of nocturnal observation first, long before the students have had a chance to chatter away their discomfort in the new place, serves three very important purposes. To begin with, it removes the triple-tasking, worrywart, urban-neuroses, monkey-mind obsessions, thrusting them into nature's here and now. It also heightens their attentiveness to their surroundings for all of their stay, instead of allowing them the normal two days to shift. Finally, the exercise introduces the students to one another in a deep, meaningful way when they read their "night watches" aloud; they are enchanted by each other's perceptiveness, which creates a sense of trust. In short, it defines a new way of being among them. I've turned many into scotophiles who will continue to enjoy the night.

All usually come back with intense, sensual, and specific writing without having to name a single plant. The days then proceed from delight in the sensual nature of the place around them, to the need to know relationships between organisms, to the need to know the specific names of plants and animals.

That process also is at the core of my mother's teaching in the Grand Canyon, where she helps adult learners begin to feel at home. She'll ask them to see if "together we can learn five rock formations, ten trees, twenty flowers and recognize five plant families, ten vertebrates, fifteen insects." Of course, some are initially resistant, but my mother knows that will change "because I've happily watched it happen trip after trip":

> After they stop focusing on absent cell phones and e-mails, become comfortable here, this canyon charms people. Most people are curious and enjoy finding out about where they are, intrigued by new places. When they begin to notice what's around them more closely, they a priori become good observers, and good observers go away with a head full of knowledge that will brighten a day, warm a winter. They will quickly learn the names of new plants, what constellations glitter on a bracingly clear cold night, recognize who made the tiny footprints on a cattail-rimmed beach, giggle at the burping of new spring red-spotted toads, take endless pictures of flowers. They will surpass those meager numbers of my list with verve and here, in this entirely new place, they will come to speaking terms with the natives of Grand Canyon. To their surprise, they will discover another home, one that will enrich and expand their knowledge of their rim world home.

Discomfort, disgust, cold feet, inclement weather, and sheer terror are teaching tools we share. Well known for her beauty and adrenaline thrills, Mother Nature does not teach by these alone. She has superior sensations in mind to carve out our humility and sense of place on a limited planet. Through torturous firsthand experiences, I aim to radicalize a student in the sense of becoming a person who is willing to give up creature comforts, part of his personal gain, for the greater order of the planet and other species.

It is not always possible to teach natural history while rafting, climbing, or mountaineering, but it is possible to keep sense impressions at the center of our teaching. In the month-long block course at Colorado College that my mother teaches, for instance, she focuses on writing and reading natural history. The students are asked to know their subject, to do accurate research, and to write in a literary style. Her reading assignments from Robert Finch and John Elder's *Nature Writing* acquaint them with the charm, communicative qualities, and important niche of natural history writing.

We both teach a wide variety of courses, from college to field institute to private workshop, and our texts differ for each. Each course uses the guidebooks for that particular region and prodigious handouts that I have developed over the years.

I don't like textbooks, which emphasize the intellectual interpretation of nature. Although the work of other nature writers, particularly writing of the area in which I teach, is wonderful, I have been forced to develop my own handouts, designed to connect students immediately to the nature around them. I hope to collect these into a book, but then again, they are specific to place and context.

Ann Zwinger's class emphasizes how to keep excellent field notes and how to develop familiarity with identification and research techniques. Upon return to the Colorado Springs campus, students must combine their fieldwork and research to create an essay that educates while it entertains. It must also acknowledge the complexities of the natural world and endow readers with a greater understanding of the world in which they live—and, above all, be beautifully written. She wants them to hear the sounds of another world, discover the wisdom of clouds and the silence of stars, and, most of all, truly enjoy becoming acquainted with and being in the natural world:

> Teaching in place is tantamount for teaching about place. When I teach "Writing the Natural History Essay" at Colorado College, students and I spend a week at CC's adjunct campus at Crestone, Colorado, where they have an intense week of note-taking, identifying plants and animals, sleeping out on the dunes and sometimes getting snow-covered, learning about place by being in place. With their firsthand field notes they can go back to the college library and learn and write about what they've seen.
>
> Teaching in the Grand Canyon and other sites with adults, it's the meandering through a patchy meadow, poking around and hiking up a cliff or wandering down a canyon that gives meaning to what's there. Learning geological formations, making a wildflower or a bird list, is so much easier out-of-doors than through a window or in a library. Showing slides on a screen is effective as reinforcement, but the initial contact has to be hands-on: the snug feel of being zipped in a rain jacket while watching raindrops pop across the river surface, sleeping under a tree hung with hundreds of cicada shells like little crisp ornaments, feeling the bite of the wind and the sound of sand singing—none of these indescribable moments can be imagined. They must take place in place, and the benefits are far beyond the immediate.
>
> Learning about place, students become familiar with something that was once strange and perhaps even threatening, and in essence they are extending their sense of home, observing the impeccable interactions of a healthy natural world, gaining a sense of stability in a world that seems intent upon tearing itself apart. It is by knowing a place that we come to

know ourselves and strengthen our connections with the world about us, and learn what's necessary to keep all of us—prairie dogs, water striders, redbud trees, mariposa lilies, water fleas, lichens, bark beetles—all that incredibly complex and complicated living world, in working order. We learn only by being there, and letting place work its magic.

Most recently, I've been applying these principles in a master's course for teachers called "Increasing Focus Through Natural History, Creative Writing, and Art."[1] Awkward as this title is, it draws desperate teachers hoping to teach students suffering from attention-deficit/hyperactivity disorder how to focus and, thus, to learn. We split our time between a large classroom and the beautiful woods on Lake Washington, north of Seattle.

The young forest that now graces Lake Washington's northeastern edge is our classroom. Even in a third-growth forest, students can walk a mile down to the water through trees draped in long epiphytes, fluted trunks of western red cedar, and massive mounds of moss over troll-shaped tree roots. They gain a keen sense that living here is not a stop-and-go proposition, as it is just down in Seattle. Here trees grow upon trees (hemlock will sprout high in a Douglas fir, then drop roots down twenty feet to the forest floor). Water trickles through in sluggish tannin stain, cedar soup from trees steeping. Even stumps, remnants of the older forests, some with chinks from old logging practices, become fascinating treasure houses of new growth, and thus new learning. Every surface is crammed with growth, growth on top of growth, mushrooms, huckleberry, tiny hemlock, slime molds, salal, amallaria, and wintergreen crowding for territory. Pan could be playing his flute inside this forest, and we would expect it.

I work them hard during our short time together: ground observations, recording saturation details, recognizing plant families, using
field guides, creating metaphors from natural shapes, learning a simple
Japanese drawing method based on meditation, speculating on the
causes of evolution in shapes and colors, and encountering the aha's
of strange adaptations. I teach them the importance of the scientific
method in citizen science (such as the Audubon Society's Christmas
bird count and the Northwest's citizens' beach watch) and to create
journals by examining role models of traditional natural history journals
(Leonardo da Vinci, Darwin, and Audubon).

This observation course takes students through a series of specific
stages, from close observation to completion of lively, scientific drawings
of the natural world. The children my students work with in public
schools are increasingly impatient, rushing through assignments with
unfocused energy and superficial observation. The Japanese method
of drawing-meditation that I teach requires slowing down, shifting
one's state of mind, and honoring the reality of the object to be drawn.
Drawing, taught as a primary subject to all Japanese children, opens
essential areas of the mind, which are the basis not only for art and
science but also for healthy human beings. In addition to drawing,
other skills of deep observation are practiced. The natural history of
our particular region (Washington and Oregon) is better understood
through drawing and writing.

At first many students are afraid to draw, but then, while attempting
to render some ordinary object, they realize how extraordinary the
natural world is. They discover the miracle inside a common daisy or the
branching of a tree, the structure and secrets of a cattail's seed puff. We
also explore the world through loupes, creating description, metaphors,

poems, and we learn the psychology of how aesthetic experience enhances observation and scientific thought.

I teach them how geological forms shape the natural and human histories of our area, and help stir their passion for the natural history of the Northwest. I also show them how to create field journals of scientific and aesthetic value, exhibiting my twenty-plus big black notebooks for inspiration. At night at home, they are required to record all their senses in their neighborhoods, often discovering natural phenomena in the place where they have lived for years. One student, who included her child in her exercise, said, "I never knew we had a stream nearby. My three-year-old heard it and was delighted. We have looked at a map to discover where it comes from and how it goes down to Puget Sound. Very cool."

Teaching nature out in nature always involves surprise. Sudden shifts in perspective, gut reactions, swift dissolves of stereotypes, and spontaneous human curiosity: these are the gifts I live for as a writing professor and as a natural history teacher. I can endlessly compose lectures, but nature's serendipitous one-upsmanship ultimately trumps my very best teaching.

One method of evoking such serendipity is a spontaneous Socratic dialogue in a specific place, using a resident organism to "ask" the students questions. The challenge: knowing a little bit about geology, biology, ecology, mycology, and ethnobotany. When hanging around in nature (unlike in my *departmentalized* university), I give myself permission to become a generalist. I simply study what my heart directs in no apparent order.

This is the lesson of Socrates' Slug.

One fine spring day, a botanist and I led public school teachers through a thick Washington forest. They had come to acquire skills that would allow them to teach the new environmental requirements for public schools. The botanist was spitting out the names of plants like a drill sergeant.

"Argh!" gasped the teachers as a slug crossed their trail.

"Ah, this fine fellow has a place within this forest. Do you know what he/she eats?" I asked, delicately picking up the creature between thumb and forefinger.

"*My garden!*" came the chorus.

"No, I mean out here in the forest."

Most Seattleites never consider a slug other than as a nuisance.

"His favorite food is stinging nettles," piped up an observant student.

"Good. What do the stinging nettles do for the slug's body?"

"Add stings."

"Excellent. Nematocysts, actually, tiny packages of stinging chemicals, which do what?"

"Aha. Keep birds from eating them." Their disgust was transforming into delight.

"Yes. Nematocysts are harmful for birds, but how do they affect larger predators, say mammals?"

"Don't know. Hmmm, probably won't hurt a big mammal, but what will happen if one does try to eat a slug?"

"Let's find out." I grinned and lightly ran my tongue along the bottom of the slug.

Again, groans of disgust.

"My tongue has gone numb from the stinging nettles. Isn't that a wonderful example of a plant-animal symbiosis?" I said, wagging my wounded *lingua.*

"So bigger predators would choose not to eat the slug because their mouths go numb," said a science teacher.

"Oooo, but the slime is just so gross," groaned another teacher.

I poked at the glistening trail through the leaves. "In this strange substance scientists discovered new chemicals that held clues for the treatment of AIDS."

The teachers packed tightly around me to take turns gently stroking the slug. I placed him tenderly in the leaves with a deep bow of thanks for letting me cause him discomfort—then turned to find all the teachers, even the most disgusted, bowed in Zen fashion toward their most honorable, slithering teacher.

NOTE

1. Credit for this course is offered through Antioch College's Seattle campus.

COURSE READINGS

Hugo, Richard. *The Triggering Town: Lectures and Essays on Poetry and Writing.* New York: Norton, 1979.

Kirk, Ruth, with Jerry Franklin. *The Olympic Rain Forest: An Ecological Web.* Seattle: University of Washington Press, 1992.

Nabhan, Gary. *Cross-pollinators: The Marriage of Science and Poetry.* Credo Series. Minneapolis: Milkweed Editions, 2004.

Zwinger, Ann. *The Nearsighted Naturalist.* Tucson: University of Arizona Press, 1998.

Zwinger, Susan. *Stalking the Ice Dragon.* Tucson: University of Arizona Press, 1991.

CHAPTER 3

Uplift and Erosion
Together Along the San Gabriel Front

BRADLEY JOHN MONSMA

SCENE ONE

It's raining. An environmental literature professor strolls from the Chandler-esque noir of the Pasadena street into the warm light of Vroman's bookstore. He's pondering once again why the books and places he loves seem so little noticed by everyone else. Suddenly he overhears a clerk trying to maintain patience with a persistent customer.

"I'm sorry, but there is no book called *Los Angeles Against the Mountains*. Nothing like that by McPhee. It's not in the computer."

"But that's what it said in the newspaper, and that's what my friends said. I know it exists!"

"Not in the computer."

Seeing an opportunity to save the day, Lit Prof interjects, "It most certainly exists, as the last chapter of *The Control of Nature*. I'll bet the computer's got that one. In fact I noticed there's one copy left on that long John McPhee shelf. It's a good read."

The customer smiles at Lit Prof with the gratitude usually reserved for firemen and superheroes. "Dude, you saved the day. I've been looking all over. My hip urban planner friends with square black glasses want me to read this. They tell me it helps make sense of all the rain and overflowing debris basins and all the weather chaos on the news."

Lit Prof smiles knowingly, confidently, like a hero or hipster, buoyant with the hope that McPhee's environmental history is helping create a community more aware of its tenuous occupation of the border between the wild and the willed.

SCENE TWO

Two weeks later. It's still raining. Lit Prof and his botanist colleague sit at the head of a classroom. The lights are down and there are images on the screen of a sunny suburban foothill neighborhood on a steep hill. One image shows an oleander hedge concealing a large debris basin just across the street from the houses.

"This is it," says Lit Prof. "This is the neighborhood McPhee writes about. This is where the families were trapped in their houses by mud choked with boulders and automobiles 'like bread dough mixed with raisins,' page 185."

The two professors show a graphic of the erratic rainfall totals of the Los Angeles basin, talk a little about the burn cycles of the brushy chaparral hillsides, and then put up a map of the area.

"You gotta to be kidding!" says one of the students. The professors smile at each other. It happens every year.

"I grew up right near there," the student predictably continues. "Nobody ever told me about this. Are you sure it really happened? Damn! My folks and my little sister still live there. Could all this shit happen again?"

Wind-driven rain slashes against the window. A few students chuckle nervously. The professors raise their eyebrows; they don't even have to ask the questions. Why do we not read the signs all around us? Why do we all so easily forget the past once it has been cleaned up and hosed off?

SCENE THREE

It's sunny and green after all the rain, a perfect spring day. The professor and the botanist and twenty students make their way up Trail Canyon, one of the drainages McPhee mentions. The creek level has come down some, but it's still flowing, and there are many sketchy crossings on rocks and logs. The nimble among them dance across; others still shake after the eighth crossing.

Somebody finds straight alder staves and passes them to a girl and a guy frozen midstream on a log. The two cross with the crutches and pass them to the next person.

Eventually the trail leaves the stream and climbs up into the chaparral. Some of the stronger hikers head up the switchbacks while a few hang back.

"We're going up *there*? You guys are trippin'!"

The groups get strung out along the trail, and the professors feel out of control, hoping no one slips at a washout or over a hundred-foot drop. They worry and mumble doubts between themselves, but the students look after each other. Somebody always waits for the slow ones and offers encouraging words. Along the way, a few find the plants they've written about earlier in the semester, and they stop to give the group impromptu field lectures. *Dudleya lanceolata, Salvia apiana, Quercus agrifolia.*

At lunch, while students compare which cell phone plans have the best coverage, the botanist takes an ice-cold swim in the pool at the base of a thirty-foot waterfall. Afterward, it's back to business, and the professors arrange the class into small groups to huddle over wildflowers and plant key books. The professors shuttle along the trail between groups as students holler questions.

"Hey! What's 'hirsute' mean?"

"That would be like Nico." Laughter.

"Okay, I get it. Is he pinnately compound, too?" Occasionally shouts of triumph go up, signaling a positive ID.

After all the stream crossings on the way back, no one seems to know how to part ways, so Lit Prof tries to say something meaningful and the botanist cracks jokes and throws a guy's shoes into a bush when he's not looking. There are only a couple of weeks left in the semester, and everyone's starting to feel that what they have created to hold them together is about to be undone. The curtain's coming down on a fine performance. That's the way it goes; they've all been through it before.

SCENE FOUR

Lit Prof sits alone in his study at home typing and retyping the same sentence, rearranging words here and there. There's a raven on the telephone pole, and the dog over the back fence barks at it incessantly, as it has for the past six hours, just like yesterday.

Flashback to one year earlier: Lit Prof approaches the dog neighbor's door with a bouquet of native wildflowers from his front yard and a basket of tomatoes.

"As the folk song says," smiles Lit Prof, "the only two things that money can't buy are love and homegrown tomatoes. Now about your dog . . ."

Lit Prof and the neighbor talk about the dog. They pet the dog and play with it. It's quiet now, a nice dog. The neighbor seems to understand about the noise, but the dog will bark for hours, then days, then weeks, whenever the neighbor leaves for work. Lit Prof will tape to the dog neighbor's door a copy of the Billy Collins poem "Another Reason Why I Don't Keep a Gun in the House." It won't help matters.

Cut to present: Lit Prof grins at the memories and at the dog still barking a year later. He's just accepted the offer of a new job and the house has just sold. Yesterday he noticed a FOR SALE sign on the dog neighbor's house, too. He retypes the sentence again and tells himself he doesn't give a damn about the dog or the neighbors.

Okay, I'll admit that I've stuck too close to the truth. These scenes might make a better film if the class hikes up Trail Canyon the day it rains five inches in an hour and a minor character—a C student—dies in a debris flow to form a lasting bond between the A and F students. Or maybe if Lit Prof really does keep a gun in the house, and the girls swim in the waterfall rather than the botanist. Good movie or not, these scenes have me thinking about how ephemeral communities—college classes and neighborhoods—assemble and disassemble. I'm caught between an ideal of constancy—that's what the great environmental writers tell me to pull for—and the reality of change. I'm glancing back and forth trying to evaluate each shift by finding good in things that don't last forever.

These issues were unavoidable for me as I taught McPhee and other writers as part of a team-taught class called "California Natural History and Nature Writing." I felt a little hypocritical talking with students about deepening our idea of home and our commitments to justice and to communities that include non-humans while interviewing for a new job and knowing I would put our house up for sale and move just far enough away to uproot my own growing commitments to a chosen place.

Teaching helped me face up to the full effects of moving on. I wasn't just moving away from a problem with a dog (I tried!) and a neighbor (I really tried!). There is a development planned for the chaparral ridge above my former north Los Angeles neighborhood where I would ride my bike, watch critters, and meet neighbors. The Texas developer wants to move whole hillsides and is seeking an exception to the zoning laws to build more houses than are currently allowable. The community organization has been impressive, a carryover from a successful fight to save another section of the Verdugo Mountains a few years ago. I was starting to get involved in this one; it was close enough to matter, to affect my quality of life, my community. Before I left, I could see some of the same people mobilizing to try to stop Home Depot from taking over the old Kmart big box and doing in the little mom-and-pop businesses like the hardware down the street owned by a Korean couple. Then there was also the little par-three golf course whose owners wanted to sell to a housing developer. I wasn't sure what to think of that one, but it was clear that these were all environmental problems, "environment" including things like the capacity of local schools, emergency response times, crowded streets, and the diversity of small local businesses as well as the need for open space and the crucial wildlife corridor between the San Gabriels and the Verdugos, a beautiful mountain island with city on all sides.

But it's hard to be involved in all that now, since I've moved away just when I had been in one place long enough to see it as home, to feel committed. Before I left, I looked up and down the street of my soon-to-be former house of six years, and realized that, of thirteen houses, seven had different residents than when I moved in. Three had turned over more than once. In only a few years, less than half a neighborhood will even remember there was ever a trail on the ridge or a hardware store you could walk into and ask a guy who knows about nails about nails. This is all about shifting baselines or an urban version of a phrase I heard somewhere: "generational ecological amnesia." Except that here the generations are much shorter than a reproductive generation. We lose memory when we move around, even when we do it for good reasons.

It was unsettling, therefore, to be talking all the while about McPhee's essay, for the place he writes about and my home were the same, and one of his central musings has to do with the loss of memory and the effect of this loss on a community's understanding of its place in ecological systems.

This is exactly why we included McPhee on the syllabus. This is why we hiked into a canyon he mentions—to fight the forgetting, to use compelling words and field experience to attach us more tightly to a place. With McPhee as a starting point, my botanist colleague explained to the class his research on chaparral fire ecology and seed germination. We talked about the hydrophobicity of soil after a fire, one of the elements, along with exceeding amounts of water, needed to produce an "event," which McPhee likens to the loading of an eighteenth-century muzzle loader: "the ramrod, the powder, the wadding, the shot. Nothing much would happen in the absence of any one component. In sequence and proportion each had to be correct" (203). The 2004–2005 cycle was a sloppy load, with record winter rain but an easy fire season the previous fall, and still people lost their lives. I helped students compare the flow of McPhee's battle metaphors with newsclip discourse of nature at its most wrathful. We showed maps of the areas McPhee mentions and recent pictures of the perfectly rebuilt neighborhood he uses for his dramatic frame of a family terrorized in their home by the inexorable and inevitable flow of liquid landscape, a "rock porridge" of mud, boulders, and automobiles.

Perhaps by learning the botany and geology of their home places, students might consider for the first time the environmental history as well. Indeed, every year at least one student has grown up near the neighborhood McPhee writes about, the one with the debris basin hidden by a neat hedge of oleander. Or they've come from other neighborhoods with other debris basins. There are plenty of them. Every year those students have no knowledge of such a thing as the battle McPhee describes between the mountains coming down and the city trying to hold them back. "A superevent in 1934? In 1938? In 1969? In 1978? Who is going to remember that?" McPhee asks. "Mountain time and city time appear to be bifocal. Even with geology functioning at such remarkably short intervals, the people have ample time to forget it" (203).

McPhee makes the observation that there has been plenty of publicity of the debris flows and the destruction they cause. He comes to the conclusion that no amount of awareness will make people truly consider the risk of living in the beautiful dangerous places they do. Drawing on John Burroughs, he writes that people "would rather defy nature than live without it" (236). And yet McPhee soldiers on. He writes anyway, presumably with the knowledge that his writing is unlikely to have any major impact.

Perhaps this is one source of McPhee's gentle irony, his bemusement at the pride and ignorant optimism with which we always seem to rebuild in the same places after entirely predictable disasters. In this year of wild storms and my own participation in the process of movement and forgetting, I find myself grateful for his understanding of the dilemmas we all face. There are plenty of environmental histories of Southern California that upbraid us for our obliviousness toward our natural (and cultural) contexts. *Ecology of Fear,* by Mike Davis, for example. Yet McPhee seems unique with his undercurrent of compassion for us all, here together for millions of reasons, trying to make a go of it, forgetful in our optimism. It's not that he ignores the idiocy, for he quotes the boosters who perpetuate the myths and make money off the innocence of newcomers; they, too, are a California tradition.

Like the cities along the San Gabriel front, our class was made up of longtime Californians and newcomers, from other states and other countries. But there was something different about the class, too. We created our usual rites of passage with field trips to Yosemite and the local mountains, and students proposed and planned a barbecue to show pictures and relive them. Everything seemed to come together. Students openly expressed concern for each other and for the professors. They helped each other deepen their thinking and improve their writing. They learned more because our "community" found a depth beyond that of the normal classroom. And the whole time, the weather was exceptional; the rainfall totals kept rising.

It occurs to me that perhaps the weather and the success of our class were related. Maybe the feeling of extraordinary times brought us together as a learning community the way disasters bring together neighborhoods and cities. We know about this in Los Angeles. Nothing sparks random acts of kindness and conversations between strangers in the City of Angels like an earthquake, a big fire, a riot, or a season of record rain. Maybe our class benefited from that effect. And that has me worried. After all, we've also experienced how quickly civility and neighborliness dissolve back into the routine of nods between neighbors and downcast eyes on the street. We're not accustomed to sustaining the relationships that seem so rewarding while they last. It's easier to finally give a grade and move on than it is to figure out how to help students truly challenge themselves to take a new step in their learning. New friends from a class quickly become old friends to be replaced by new faces next semester. Could it be that classroom learning communities mimic, or even teach, the disassembling so common to our neighborhoods, cities, and landscapes? Academic rhythms encourage intense new relationships, yet they also regularly break apart what has come together. Often the breaking happens just before the real and difficult work would begin—the practice of the theory. It all reminds me of the geology we live with here: "The San Gabriels, in their state of tectonic youth, are rising as rapidly as any range on earth. . . . Shedding, spalling, self-destructing, they are disintegrating at a rate that is also among the fastest in the world" (McPhee 184).

As hurricanes, floods, and tsunamis make clear, it's not just at the edges of Los Angeles where these dilemmas present themselves. Perhaps there will soon be a whole generation of environmental educators and students for whom the violent reshaping of landscapes and communities is the norm. Perhaps in these new contexts, the brevity inherent to education will seem to be more a model of sustained relationship. And if uplift and erosion is the hand we are dealt, I can't think of anything to do but to continue to use field experience and collaborative learning to make the communities of our classrooms meaningful and intense, maybe even more intense for being brief. There is good in that while it lasts.

⟡

I always feel better having introduced McPhee to students or helped someone search him out in a bookstore. There are quite a few people in the Los Angeles basin and foothills, and this is a small start, but those who read "Los Angeles Against the Mountains" might understand for the first time that their landscaping rocks arrived in their front yards by geologic processes that are still active. They might look up at Mount Lukens or Mount Wilson, red at sunset, with new wonder, expectation, and caution. This is the kind of text that inspires participatory, active reading, and as such it is a performance that gets people out to walk where McPhee walked, reading the age of the chaparral, gauging the intensity of the next fire, and considering the shape and trajectory of the watershed. McPhee may be a storyteller who helps some people, including those in our class, to see their city for what it is and to participate in the ongoing negotiations along its dynamic border between the urban and the wild.

Furthermore, McPhee isn't the only way the story gets told. Some of those who don't find their way to McPhee's essay will wander over to Bolton Hall in Tujunga, an old house made of foothill stones, where white-haired matrons tending to the local past will show them pictures of the 1969 flood that took out the graveyard and sent coffins down Summitrose Street and into people's backyards. A front yard in Sunland filled with native plants might persist from one owner to the next, offering a different narrative than that of lawn grass and imported water.

The woman who lives up on the ridge will continue to stop mountain bikers to hand them flyers for the next community meeting.

This is what I wanted my students to feel and understand from their reading and their walking—that we might have patience and compassion for all of us tasked with building communities of memory. We know from our own lives that it's not easy, after all. Perhaps our class, though temporary, may be a performance that opens us to the possibility to connection and renews us for the task ahead. Like all performances, it is ephemeral but transformative.

FINAL SCENE

On the last days of class, students read parts of their essays of place and narrate for the class the ups and downs of the writing process. Independently and unprompted, the writers focus less on place than on their relationships with others in places. They identify similar moments as the start of the class, coming into something different, and more, than a class. Of course they rhapsodize some about Yosemite's dramatic vistas or seeing John Muir's favorite bird, the ouzel, dipping in and out of the Merced River. But more, they are taken by the moment when the class went off-trail at night among the boulders and roots to approach the base of Yosemite Falls. The darkness and the fear many of them felt compelled them to give and receive help from each other, moving from rock to rock, holding back bay tree branches for the next person to pass. New friendships began, and small groups of friends opened to others. Some mention that the hike in Trail Canyon closer to campus near the end of the semester—helping each other cross streams—confirmed and reinforced the cohesion of the class and individual relationships within it.

As Lit Prof and the botanist listen, they realize that most of these moments had passed unnoticed by them as they worried about people slipping and breaking ankles. When Lit Prof had asked everyone to be silent for a few minutes and to attend to the sensual richness below the falls—the icy spray, the starlight filtering through high clouds, the sound of the windblown water, the scent of wet granite and bay leaves—he spent his silence worrying about who might be cold and uncomfortable, about how much silence a group of young people could take. Lit Prof begins to understand that his role in the performance of the class had hindered a certain type of participation. Even as the whole class was forming a foundation of trust and respect for the rest of the semester, Lit Prof was largely unaware and on the outside, worrying about the script, the props, and the lighting.

Lit Prof, however, begins to understand the communal dimension of the students' learning when he hears them reading their own writing and daring to express how their initial fear and uneasiness grew toward confidence through the help of others and through knowledge gained by reading authors like John McPhee.

As he listens to students read and talk, Lit Prof thinks of Richard Bauman's classic, *Verbal Art as Performance:* "It is part of the essence of performance that it offers to the participants a special enhancement of experience, bringing with it a heightened intensity of communicative interaction which binds audience to the performer" (43). Lit Prof feels himself part of an audience now and realizes that the reflective writing has become what Bauman would call the "metacommunication" of the performance. It has allowed students to take on the role of performers whose words elicit the participation and energy of the audience, and with that the potential for transformation, not just of themselves but of the world around them as well.

It's the last day of class, but it has been a good show, thinks Lit Prof. If the curtain comes down as everyone applauds to shouts of "Brava! Bravo!" well, that's just what happens on the stage. Performers and audience alike emerge into the light or night of a real place and see new possibilities for it and for their relationship to it—or wherever they find themselves in the future. That's enough.

Exuent

COURSE READINGS

Gilbar, Steven, ed. *Natural State: A Literary Anthology of California Nature Writing.*
Jeffers, Robinson. Assorted poems.
Lopez, Barry. "A Scary Abundance of Water." (This essay on Lopez's childhood in the San Fernando Valley and the history of the Los Angeles River was published in *LA Weekly.*)
McPhee, John. *The Control of Nature.*
Monsma, Bradley John. *The Sespe Wild: Southern California's Last Free River.*
Muir, John. *The Mountains of California.*
Schoenherr, Alan A. *A Natural History of California.*
Solnit, Rebecca. *Savage Dreams: A Journey Into the Landscape Wars of the American West.*

WORKS CITED

Bauman, Richard. *Verbal Art as Performance.* Prospect Heights, Ill.: Waveland Press, 1977.
McPhee, John. *The Control of Nature.* New York: Farrar, Straus, and Giroux, 1989.

CHAPTER 4

A Teacher on the Long Trail

JOHN ELDER

During my four years at Middlebury College and the Bread Loaf School of English, I've often taught courses whose titles recycled combinations of the words "nature," "writing," "landscape," "reading," "place," "Vermont," "mountains," "watershed," and "home." Looking back now at this comical continuum, however, I can discern a progression in it—a story with two main chapters, and with a third just beginning. The one that dominated my first couple of decades of college teaching combined equal parts of hiking and of reading and teaching nature writing; it culminated with my request for a split appointment between English (my original department at Middlebury) and Environmental Studies. The subsequent turning in my path as a teacher began about a dozen years ago, with a series of expedition-related classes at Bread Loaf, my summer institution. These outings built upon my individual hiking experiences through extended camping trips with my classes, while they complemented the previous study of nature writing with more focused attention to natural history and Native American literature.

More recently, I am beginning to focus on issues of conservation, community, and sustainable communities. In none of these regards have I been a pioneer, by any means, and in all of them work is now being done by others that goes far beyond what I have accomplished. But though mine might be, as Wordsworth says in framing *Michael*, "a history homely and rude," I would like to relate it as one teacher's path through the thickets of academia and into the mountains of home.

Each of the chapters in my story as a teacher begins on the Long Trail. This is the footpath leading from the Massachusetts border to Canada that coincides with the Appalachian Trail in its southern portion. During our first several years in Vermont my wife, Rita, and I spent quite a few weekends on the Long Trail, which passes just a few miles to the east of Middlebury. Having grown up in the Bay Area, where my own earliest hiking experiences were in Yosemite and hers were in the mountains above Santa Cruz, we found that our initial impressions of the Vermont mountains were not altogether positive ones. Whereas California summers were predictably both rain-free and mosquito-less, here we contended with muddy trails as we swatted away black flies in June as well as mosquitoes in the latter part of the season. These mountains were also largely devoid of the sublime vistas of the West; the Greens are a rolling and heavily glaciated range, and only a few summits on the Long Trail even rise above the tree line. But hiking along with our eyes on the ground and the nearby trees, we slowly became attuned to the distinctive beauties of northern New England. These are moist woods. Several years of leafage composts sweetly in the filtered sunlight of a morning in summer, beside trails edged with sorrel, bunchberry, and blue-bead lily. Grand glacial erratics punctuate the forests of Vermont. The larger and flatter of such boulders are often carpeted with the fern known around here as rock-tripe. The smaller, rounder ones often have a yellow birch balanced on top, its thick roots roping down on either side to find purchase in the scanty, black mountain soil. Majestic bracket fungi rise stairstep fashion on funky snags that are also likely to hold pileated-woodpecker holes as big as a couple of volumes of the *Encyclopedia Britannica*. In hiking the lower elevations of the trail we would sometimes hear barred owls calling at dusk, while when arriving at the highest ridges we would usually be greeted by the piercing call of the white-throated sparrow.

The pleasures of such a forest are intimate and familial, not only because they are often small and have to be concentrated on to be fully appreciated, but also because they are limited in variety. In the forests of Costa Rica we might have encountered hundreds of species of trees and birds during several days of hiking. Even in the southern Appalachians we would have seen much greater ecological diversity than here. But in Vermont, once you've learned twelve or fifteen trees you can go for hours without seeing a new species, and the beauties of rock, moss, fern, and birdsong are endlessly repeated—variations on a theme whose primary elements are deeply familiar. The pleasures of these woods have less to do with glamour and novelty than with the satisfactions of a beloved home.

Coming to identify with such a landscape amplified my appreciation for nature writing in the Thoreauvian mode. I'd loved Thoreau since my parents gave me a copy of *Walden* for my fourteenth birthday in Mill Valley, California. But putting down roots in Vermont made not only that book but also "Walking" and the journals landmarks in my life. Thoreau's celebrated sentence from "Walking," "In wildness is the preservation of the world," describes a quality of alertness that is at a premium in the modest geological scale and second-growth woods of New England. My renewed relish in reading Thoreau gave me an appetite for exploring the genre as a whole, and I soon began teaching authors like John Muir, Mary Austin, Rachel Carson, Aldo Leopold, Edward Abbey, Annie Dillard, Barry Lopez, and Terry Tempest Williams on a regular basis.

Poetry became increasingly central to my classes, too, since Robert Frost's writing proved equally helpful to Thoreau's in giving me a fix on the landscapes around our home. What a gift it is to live in a place whose landforms, cycles, and history have been incorporated so specifically into the work of a great poet. Focusing on Frost in this connection made me both look back to Wordsworth and the Romantic revolution and immerse myself in the work of such contemporaries as Gary Snyder, Seamus Heaney, and Mary Oliver. I began regularly teaching a course called "Visions of Nature" that featured many of the nature writers and poets I've just listed and that (with the college's blessing) replaced my lecture class on modernism. This new offering ended up becoming the humanities gateway course for our environmental studies major and strengthening my connection with that program.

While my teaching was shifting as has just been described, Rita and I were also discovering that we wanted Vermont to be our lifelong home. The reading I had been doing in and out of class encouraged me to explore this homing process within the context of journaling—a practice so close to the surface of much nature writing from Thoreau to the present. Accordingly, I also regularly began to ask my students to keep journals and to develop reflective and narrative essays out of their most promising entries. Collections of essays became more prominent in the syllabi at this point, since they were so useful to students following such an observation-based approach in their own writing. Thus, whereas I might earlier have assigned *Pilgrim at Tinker Creek* or *Arctic Dreams,* I began more often to ask students to read Dillard's *Teaching a Stone to Talk* or Lopez's *Crossing Open Ground.* Pieces like "Living like Weasels" and "Total Eclipse" from the former offered structures that helped students consider their own formal options, while "Landscape and Narrative" and "Children in the Woods" from the latter addressed questions about authenticity in writing and about the complexities of naming that often came up in our class discussions.

This combination of forays into the Vermont landscape, readings from nature writers, and an energetic journal practice was so fruitful for my students and me that I wanted to develop it further. Most summers, I teach at the Bread Loaf School of English, a graduate school in literature and writing run by Middlebury College. Bread Loaf students, most of whom are high school English teachers during the regular academic year, attend for five summers to earn their master's degrees. In a given summer they can either come to the Vermont mountain campus in Ripton—at a Victorian resort about ten minutes' drive from Middlebury—or choose to study at one of our campuses in Oxford, Juneau, Santa Fe, or Asheville. The normal course load for a six-week Bread Loaf term is two five-credit courses taught by each instructor and two enrolled in by each student, with each class meeting for five hours a week. Jim Maddox, the director, gave me permission to develop off-campus courses that lasted for just three weeks but were so intense that the overall number of hours met and writing accomplished were fully equivalent to those of a standard course at Bread Loaf.

The first of these experiments, which I offered in the summer of 1994, was a seminar for ten students that combined reading, writing, and nature study and that was held entirely on the Long Trail. "Writing in the Mountains" began at the Canadian border. Over the following weeks we made our way slowly south on foot, concluding the course when we walked down out of the mountains and arrived, grimy but elated, at the Bread Loaf campus. This course was a thrilling departure for me as a teacher. Living together for these weeks—experiencing rugged terrain, exhaustion, and a hellacious thunderstorm as well as sunrise, birdsong, and innumerable moments of quiet beauty along the trail—generated a sense of intellectual community in our class beyond anything I'd ever known. People walked on their own during the day, pausing to read and write whenever they felt so inclined. When, in the late afternoons and the evenings, we gathered for discussions of the readings and workshops on our own writing, the excitement of the conversations was palpable, the comments remarkably trenchant and constructive. None of our meetings were on the clock, and as strenuous as the days' miles sometimes were, the number of hours devoted to such meetings was enormous. The work of the word and the adventure of the trail became complementary dimensions of a personal and collective quest.

And the days, especially the first several just after we set off bravely from the Canadian line, really could be pretty taxing. There's a lot of up and down in this initial segment of the trail, with some hands-and-feet work to make it over boulders on the route while also wearing heavy packs. Though I'd advertised the course as being for experienced hikers and required a doctor's letter from all applicants, everyone was tired at the beginning, and a couple of members of our group were really hurting. On the third day one student just lay down on the trail and told us to leave her there. We were of course not inclined to do so, and a couple of sympathetic classmates hunkered down to give her a snack and some encouragement. I'll never know whether when she got back to her feet and shouldered her pack it was because her energy was restored or because her colleagues had just been offering her spoonfuls of leftover hummus from a sogged Baggie that tasted as repulsive as it looked. A second member of the class found that, with the additional weight of his backpack, his ankles quickly became weak and swollen. He needed to cut a second walking stick and to go very gingerly. But both of these folks gained strength week by week. They ended up taking tremendous satisfaction in completing such a challenging hike and were celebrated for their accomplishments by the group. The woman who had needed to be urged on by hummus ended up being known by the defiant new trail name she adopted: *Towanda.*

This field-based class also integrated some kinds of readings I had not previously assigned. For one thing, this was my first real attempt to pursue natural history in some systematic way. We brought along an excellent guide to the geology and ecology of our terrain, *The Nature of Vermont,* by Charles Johnson, and divvied up individual field guides to the trees, flowers, birds and wildlife, moss, and ferns among our backpacks. (In similar efforts to minimize the weight of books in our packs, we either did the readings before setting out and brought along our notes or brought single volumes and photocopies that people passed around.) Beyond our selections from Thoreau, Frost, and a selection of contemporary nature writers and poets, we also read and discussed a collection of traditional Abenaki stories translated by Joseph Bruchac.

The Western Abenaki are the indigenous people of Vermont, and their creation tales of Gluskabe accounted in wonderfully specific ways for the glacial erratics beside our trails as well as for the northern hardwood forests through which we passed in the lower and middle elevations. They also told us about Azkaban the Raccoon, who was the Abenaki Trickster figure just as Coyote was for the native people of the American Southwest. Scholars of American nature writing have long noticed that the genre was heavily tilted toward writing by white Americans who often tended, as well, to favor sublime western landscapes. Bruchac's Abenaki tales were especially welcome both as offering a different cultural perspective on nature and as referring to the particular ecology and landforms of our own region of the country. To place the Gluskabe stories in context, we also read Leslie Marmon Silko's essay "Landscape, History, and the Pueblo Imagination," which remains the best piece I know for highlighting the differences between Native American understandings of culture and nature and prevalent assumptions in the wilderness movement.

When we walked onto the Bread Loaf campus, shucked off our boots, and dove into Johnson Pond to sluice away the dust and mud of the Long Trail, our group was both delighted to be back in the land of dry socks and hot suppers cooked by someone else and bereft at the thought of splitting up. We had enjoyed a remarkably intense experience of community, with reading, writing, and walking at its core. After we had said our good-byes and everyone else had returned to their homes in other parts of the country, I, as a local, continued to frequent the campus during the final three weeks of the term. The theatrical production that summer was *As You Like It,* and the director, Alan MacVey, offered me the bit part of Jacques du Boys, "second son of old Sir Rowland." This previously unseen brother to Orland and Oliver appears only at the end of the play, bearing the news of Duke Frederick's conversion to a religious life just as he was approaching the forest encampment of Duke Senior with a menacing army. Specifically, Alan directed me to climb up onto the stage wearing my grubby trail clothes and my battered old Kelty pack; I was to be initially oblivious to all the other characters, flapping myself over the hat with an upside-down topo map in a vain attempt to discourage the mosquitoes and no-see-ums. This broad allusion to the Long Trail course was just one of many inside jokes favored at Bread Loaf, and both the audience and I enjoyed it. But it was also a bittersweet reminder of one of the most valuable experiences I'd ever had as a teacher—a time when learning, community, and adventure were inseparable, and when I felt, day after day, that I was at the heart of my vocation.

In subsequent summers I have built on this Long Trail experience in various ways. The following year a class spent three weeks at Camp Peggy O'Brien, an Adirondack Mountain Club cabin in the vicinity of Johns Brook Lodge. We still spent many of our waking hours hiking, but had a warm, dry shelter to return to every night, a kitchen with a propane stove for cooking our dinners, and no need to carry heavy packs with us during the day. My goal was to avoid the physical exhaustion that occasionally crept into the Long Trail weeks, and in this regard the new approach was successful. In addition, though the Adirondacks are taller, more dramatic mountains than the Greens, the flora and fauna are quite similar. As Bill McKibben has pointed out recently in *Wandering Home,* these two mountain ranges define and belong to the Champlain Bioregion, which has Lake Champlain at its center and which also has many similarities of cultural history between the New York and Vermont sides. Another advantage of having a base camp was that we could invite experts to hike in and join us for an evening. In addition to meeting with Adirondack-based writers like Bill McKibben and Sue Halpern, we profited from having the naturalist Alicia Daniel, of the University of Vermont's Field Naturalist Program, spend a long session with us on the botany and geology of the Adirondacks.

This workshop with Alicia confirmed the value of augmenting our natural history readings, and our pleasure in doing tree and flower identifications in our journals, with more concentrated exposure to science. In subsequent summers at Bread Loaf I've invited her to meet with classes again as well, and I've also brought in Tom Wessels, whose remarkable guide to ecological history *Reading the Forested Landscape* I frequently include in the syllabus. A primary goal of inviting people like Alicia and Tom was to bring more sophistication about evolutionary dynamics and ecosystem analysis to our writing and reading, in order to avoid the dilettantism that is always a risk in interdisciplinary pedagogies. In the summers of 1998 and 1999 I was asked to teach for Bread Loaf at the campus in Alaska, and there too our camping and sea-kayaking experiences became much more meaningful because of the guidance of Richard Carstensen, a preeminent naturalist and artist of Southeast Alaska. Back in Vermont over the past five summers, I have begun a teaching exchange with Glenn Adelson, an ecologist who teaches natural history seminars at Harvard; I meet with his classes to talk about poetry (especially that of Frost), and he offers workshops in the field for my Bread Loaf seminars. In our mutual excitement about the way in which science can illuminate the ecosystem of meanings in a poem like Frost's "Spring Pools," Glenn and I have recently undertaken a collaborative writing project that promises to shape our teaching further.

I have one last comment about camping-based pedagogy before touching briefly on the new phase of my teaching. This is that teaching in such a setting, though memorable and rewarding, can be utterly exhausting for an instructor. When a class is so far from other support, students' health, comfort, safety, and psychological well-being are uppermost in a teacher's mind. I also remain acutely aware of being neither a trained wilderness guide nor a medic, and of thus needing to be especially careful. In Vermont I still feel comfortable enough in the terrain and confident enough in my skill level to run field-based courses on my own; for my two summers in southeastern Alaska, with its frigid waters and plentiful brown bears, I asked Bread Loaf to supply us with professional guides. Even closer to home, though, camping with students can mean a sequence of twenty-hour days. For this reason, I have not taught an expedition-based course every summer, and for the past several summers at Bread Loaf have based my classes on campus with occasional outings in the vicinity. Starting again next year, though, as I shift to teaching only in the fall semesters at Middlebury, I hope to be able to gear up for some more summer sessions in the wild.

Meanwhile, a new angle has come into my classes both at the college and at Bread Loaf. Much of the Long Trail runs through the Green Mountain National Forest, and several stretches of it, including the one just north of Bread Loaf, are in federally protected wilderness areas. I've long been an enthusiastic advocate of expanding the acreage in Vermont that is protected by designation as wilderness. The fact remains, though, that by far the greater portion of this forested landscape (including some of the Long Trail) is in little private holdings like our own family's 142-acre sugarbush in Starksboro. Such a situation makes Vermont much different from many western states, where a major portion of forest ownership is often public and much of the private ownership is by huge multinational corporations like Weyerhaeuser and Plum Creek.

Selling raw logs is such an unprofitable experience for smallholders here in Vermont that it is becoming unfortunately common for landowners to clear-cut and then subdivide their land. If allowed to continue, this trend will lead to the fragmentation of our state's forests and the destruction of much habitat for wildlife.

A viable approach to forest conservation in this part of the country requires a strategy that combines ecological goals with economic incentives, environmental education, and collaborative, community-based approaches to value-adding. Accordingly, I find myself increasingly interested in studying and teaching books—from George Perkins Marsh's 1864 masterpiece *Man and Nature* to the essays of contemporaries like Wes Jackson and Wendell Berry—that treat sustainability within a practical and social context. I focus more as a teacher on the health of our local towns, as well as on that of the forests and wildlife that surround them. Field trips to visit value-adding projects like sugarbushes and wood-fired power generators and to talk with producers of green-certified flooring begin to complement camping trips and hikes in the local wilderness areas. Just as my first thinking about ecology did, these attempts to think harder about economic and political issues take me out of my depth; it would be possible to characterize my entire development as a teacher as a stumbling foray in the direction of complete incompetence. It will be more essential than ever to invite experts into my classes. Still, such an emphasis on civic and practical matters does feel like a necessary balancing of the exploration and personal development at the heart of previous classes. It offers another way in which a path through the mountains can lead toward home.

COURSE READINGS
Models for Writing on the Trail

Dillard, Annie. "Living like Weasels" and "Total Eclipse," from *Teaching a Stone to Talk.*

Lopez, Barry. "Landscape and Narrative" and "Children in the Woods," from *Crossing Open Ground.*

Thoreau, H. D. *Walking, The Maine Woods,* and "Walking."

Guides for Drawing on the Trail

Leslie, Clare Walker. *The Art of Field Sketching and Nature Drawing.*

Books Frequently Discussed in the Field

Abbey, Edward. *Desert Solitaire.*

Austin, Mary. *The Land of Little Rain.*

Basho. *Narrow Road to the Interior.*

Carson, Rachel. *Silent Spring.*

Dillard, Annie. *Pilgrimage to Pilgrim Creek.*

Faulkner, William. *Go Down, Moses.*

Hinton, David, trans. *Mountain Home: The Wilderness Poetry of Ancient China.*

Leopold, Aldo. *A Sand County Almanac.*

Lopez, Barry. *Arctic Dreams.*

Marsh, George Perkins. *Man and Nature.*

McKibben, Bill. *Wandering Home.*

Muir, John. *The Mountains of California.*

Oliver, Mary. *House of Light.*

Sanders, Scott. *A Personal History of Awe.*

Silko, Leslie Marmon. *Ceremony.*

Snyder, Gary. *The Practice of the Wild.*

———. *Turtle Island.*

Williams, Terry Tempest. *Refuge.*

Wordsworth, William. *Selected Poems and Prefaces.*

Abenaki Stories

Bruchac, Joseph. *The Faithful Hunter* and *The Wind Eagle.*

Forest Ecology and Field Guides

Newcomb's Guide to Wildflowers.
Peterson's Guide to Eastern Birds.
Wessels, Tom. *Reading the Forested Landscape.*

Personal Projects Mentioned in Essay

"Frost's Ecosystem of Meanings in 'Spring Pools,'" with Glenn Adelson. *ISLE* Summer
    2006.
Pilgrimage to Vallombrosa: From Vermont to Italy in the Footsteps of George Perkins Marsh.
    Charlottesville: University of Virginia Press, 2006.

PART II   MAKING CONNECTIONS

CHAPTER 5

Thinking About Women in Place

CHERYLL GLOTFELTY

Creating healthy communities requires the kind of work that was once labeled women's work, work that kept women in their proper "place." Baking casseroles for potlucks—checking on a sick neighbor—soothing hurt feelings—caring for children—planting flowers—volunteering— cleaning up. This essay describes a course titled "Women and Literature" that I use to raise feminist issues related to place and community. While women's liberation can require cutting ties and casting off obligations in the name of self-determination, wholesome communities require commitment and even self-sacrifice. This literature course studies a diverse array of American women writers whose work negotiates the tensions among individual freedom, community membership, and commitment to place.

At the University of Nevada, Reno, where I teach, "Women and Literature" is offered at the senior level, cross-listed in English and women's studies and open to all majors. It fulfills the university's cap- stone and diversity course requirements. Because the course kills two requirements with one blow, it enrolls heavily, drawing students from across campus, some of whom have little interest in the subject, some of whom feel threatened by feminism. In the past I've taught the course as an anthology-based history of women in literature. While that version raises students' gender consciousness and exposes them to important works in the literary tradition, it feels a bit canned and canonical, the chronological approach tending to render the topic academic. Moreover, as my students have pointed out, the course is depressing because the literary works that are collected in anthologies of women's literature frequently end in suicide or madness, giving more time to victimization than to empowerment.

When the editors of the present volume invited me to submit a proposal for "teaching about place," it occurred to me that a place-based theme for a women and literature course might engage students at a personal level. It also struck me that women's studies courses usually focus on identity and politics at the expense of place and, conversely, that most place-based courses employ a critical approach that is gender blind. Introducing the theme of place into a women's literature course expands the purview of both feminism and place studies, offers male students a point of connection, and encourages students to examine their own relationship to place. One student's anonymous review suggests that the aspect of *place* did breathe new life into this standard course: "I came into this class apprehensively expecting a commonplace survey of women's role in the history of literature. Therefore, I was greatly surprised (and greatly happy) to see a variation on the course: Women in literature in relation to place. This topic—two seemingly separate ideas—provided a frame for strong, intellectual thinking, which I enjoyed."

In the first week of class, I explain key concepts from feminism and bioregionalism, providing a theoretical lens through which we read the creative works. We compare conceptions of home in an excerpt from Betty Friedan's *The Feminine Mystique* and Gary Snyder's essay "The Place, the Region, and the Commons." For Friedan, home means house, implying housework and housewife. In a few strokes, Friedan paints a picture of the round of chores that consumes a typical day in the life of middle-class suburban mothers of the 1950s. Isn't there something more? Is this all there is? they ask in quiet desperation that Friedan labels "the problem that has no name." While Friedan considers the experience of the homemaker, Snyder is interested in human maturation and sense of belonging. For him, home is symbolized by the hearth, where families come together, break bread, and share stories that anchor them in the world. From this foundational sense of place, children gradually identify with an expanding circle—family, community, bioregion. For Snyder, the ideal is to embrace one's bioregion *as* home, a circle of kinship that encompasses flora and fauna as well as aunts and uncles. For Friedan, then, self-in-place translates to woman-in-house, causing estrangement from the wider world, whereas for Snyder self-in-place is a liberating term, synonymous with person-in-nature, signaling awareness of the individual's membership in larger networks.

Notions of home thus emerge very quickly as fundamentally important to both feminism and bioregionalism. An effective icebreaker is to have everyone answer the question "Where do you consider to be home and why?" It is astonishing to hear the complex answers to this apparently simple question. The question dogged a few of my students all semester. "Home" may be a particularly problematic concept for Nevadans. As the fastest-growing state in the nation for the past two decades, Nevada is a mixture of natives, who have witnessed massive changes to their home, and recent transplants, most of whom have also moved several times *within* the state as their financial situation has changed. Given such high mobility and real estate turnover rates, communities are tenuous, always changing. Furthermore, Nevada, which historically has made money from liberal divorce laws, continues to have the highest divorce rate in the nation. Approximately half of my students come from broken homes, making it troublesome for them not only to identify where home is but to identify with the sense of security and family that the concept of home is supposed to evoke.

Nevada's legalization of gambling in 1931 set the state on a course where tourism quickly became the leading industry, such that today more than 50 million people per year visit Nevada, compared to its population of 2.5 million. Thus the vast majority of the nation experiences Nevada not as home but as an adult playground, a getaway *from* home. Nevada's state song, "'Home' Means Nevada," strikes one as a defensive rhetorical move to assure residents that despite the odds (to employ a homegrown figure of speech!) Nevada can be home.

While it is easy to lament Nevada's relative lack of stable communities, students seem to appreciate the libertarian ethos that attends the state's fluid demographics. "Community" has become a buzzword in bioregionalist discourse. To consider the potentially oppressive force of community on the individual stimulates critical thinking and helps us avoid a monological approach to the topic of women in place. Kate Chopin's *The Awakening* works well to problematize community. I assign excerpts from Catharine Beecher's *A Treatise on Domestic Economy* (1841), which defends the "separate spheres" ideology that deems a woman's proper place to be the domestic sphere. Barbara Welter's essay "The Cult of True Womanhood: 1820–1860" identifies the traits of an ideal woman as promulgated in popular periodicals of the era, namely, piety, purity, submissiveness, and domesticity.

With this historical context established, students readily perceive how radical Chopin's protagonist Edna Pontellier is. As the story unfolds, she violates every one of the cherished ideals for women. Students sympathize with Edna, feeling the oppressiveness of her situation, in which social norms and, indeed, the tightly knit community itself imprison her in a net of expectations as confining as the birdcage pictured on the book cover. Because students readily identify with Edna, *The Awakening* effectively dramatizes the stifling aspects of communities when they demand conformity. At the same time, the contrast between Edna and the ideal "mother-woman" Madame Ratignolle prompts us to think about a personal orientation of selfishness versus one of selflessness, a theme that persists throughout the course.

Sarah Orne Jewett's *The Country of the Pointed Firs* offers a nice contrast to *The Awakening*. Both stem from the turn-of-the-century regionalist tradition, use the sea as a symbol, depict close-knit communities, and focus on the experience of an outsider who is distinctly aware of her difference. But whereas Edna feels ever more out of place, the unnamed narrator of Jewett's novel grows ever more intimate with the places and people of Dunnet Landing. She expects this quaint Maine village to be a writer's retreat, affording her the solitude needed to complete a project, but instead she finds herself drawn to its people, more interested in winning their trust and friendship—in belonging—than in protecting her own privacy. The narrator befriends her landlady, Mrs. Todd, an herbalist who knows exactly when and precisely where to gather a pharmacopoeia of plants, how to prepare them, and what healing words to use as she dispenses them to ailing neighbors. Mrs. Todd is a character type that recurs in many of the works, an elderly woman whose strength of character makes her seem larger than life and who serves as the linchpin of the community. Given that elderly women are perhaps the most overlooked segment of the population—"invisible" is how older women frequently describe themselves—it is daring for authors to make them the most important characters of their stories.

I will admit that many students find *The Country of the Pointed Firs* dull. It seems to lack plot and narrative drive. "Nothing happens," is a common complaint. Ursula K. Le Guin's essay "The Carrier Bag Theory of Fiction" aids in appreciating Jewett's work, as Le Guin contrasts male modes of storytelling—patterned after hero-on-a-hunt tales—with female modes, metaphorically correlated to seed-gathering activities, envisioning stories as containers. Jewett's novel "contains" sketches of a region and its people, capturing their likeness before they pass away. Her novel teaches patience, and she shows us how sense of place and sense of time are intertwined. Imagine a life where one could happily spend an entire afternoon sitting with an old sea captain as he knits and talks about how much he misses his deceased wife. Time to listen, patience to enjoy a meandering conversation, time to be quietly together. These are things foreign to the fast-paced, distraction-filled, multitasking, amped-up life of the average college student . . . and professor. Dunnet Landing—and Jewett's book—can be a refuge.

Regionalist fiction such as Jewett's so beautifully renders country life that critics have tended to associate "sense of place" with the rural and natural landscapes of literary regionalism. *Bread Givers,* by Anzia Yezierska, deconstructs this common conflation of regional with rural as it vividly creates an urban sense of place, primarily by describing the soundscape of a Jewish ghetto in New York City, recording Yiddish exclamations in everyday speech and the cries of fish peddlers and sidewalk vendors. A rediscovered feminist classic, *Bread Givers* tells the story of Sara Smolinsky, fourth daughter of Old World parents, who expect her to be a dutiful daughter, one who will obey her tyrannical father, submit to an arranged marriage, and help support them in their old age. Sara, however, has "American" dreams of getting an education and escaping the grinding poverty that plagued her girlhood. She talks back to her father, runs away from home, works her way through school, gets a scholarship to an out-of-state college, and becomes qualified to be a teacher. What is interesting in the context of the course is how out of place Sara feels in the picturesque college town. The town is just the kind of place she had fantasized about—clean, peaceful, and green— but, once there, she feels rebuffed and alienated by her difference from her more privileged, more American, peers. After graduating, Sara obeys a homing instinct and returns to New York to teach in a school near Hester Street, the squalid neighborhood from which she had fled.

A woman's quest to find a place that suits her frequently intertwines with a quest to find the right mate. Zora Neale Hurston's *Their Eyes Were Watching God* explores this twofold search for the right fit. Janie Crawford's first husband takes her to his isolated farm, where she feels overworked and lonely. She soon elopes with Joe Starks, who erects a general store and becomes mayor of the newly incorporated all-black town of Eatonville, Florida. Joe's store being the social hub of Eatonville, Janie is now at the center of the community. The problem is that Joe—the "big voice"—expects Janie to be silent and proper, deeming it unseemly for her to lower herself to common porch talk. With Joe, she is cut off from community not by geography but by gender expectations of the propertied class. When Joe dies, Janie marries a happy-go-lucky man named Tea Cake, who takes her down "on de muck" to the Everglades, where he finds seasonal work picking crops. Although Tea Cake does not expect it of her, Janie chooses to work in the fields as well, and here she seems happiest, working alongside friends by day and partying with them at night. From a feminist perspective *Their Eyes* is conservative, in that Janie does not "make herself" like Sara Smolinsky does. Rather, she allows her life to be shaped by the men she marries. However, similar themes of women coming to voice emerge at key moments in the novel, when Janie talks back to Joe Starks or speaks up in public. From a bioregional perspective, Janie's mobility is noteworthy, as each place she lives enacts a different relationship to community, from being cut off from community on a farm, to being above it in Eatonville, to being part of it in the Glades.

Themes of mobility are taken to an extreme in *Thelma and Louise*, the Ridley Scott film about two women on a wild road trip. When I assigned this film, I recalled it as a freewheeling story of two women who had made the big break from confining lives of domesticity. In my memory two friends went joy-riding in a convertible under the spacious skies of the American West. To my surprise, re-watching the film with my class, I discovered that the trip was actually one of evading the law, Louise having killed a man who was about to rape Thelma during what was intended to be just a weekend excursion. The trip gets desperate and even suicidal as the law catches up, but ironically one's lingering impression is of two women relishing the freedom of the road, having escaped the controlling men in their lives. For them, being home means being stuck and dominated, while being on the road offers an exhilarating sense of adventure and freedom from commitment, traditionally a male prerogative, here inverted to leave men in the dust. *Thelma and Louise* is a provocative film to use in a course on bioregionalism because it challenges values of rootedness and home, opening up a space to talk about the joys and risks of travel.

An ideal book to contrast with *Thelma and Louise* is one that speaks of love and loyalty to the students' own region. For a class at the University of Nevada, Reno, the perfect book to bring bioregionalism home is a recent essay collection titled *Sharing Fencelines: Three Friends Write from Nevada's Sagebrush Corner,* by Linda Hussa, Sophie Sheppard, and Carolyn Dufurrena. Each of the authors comes from an urban area but falls in love with the sparsely populated high desert country of the Great Basin. Each makes a commitment to stay. Their personal essays bear witness to their deep connection to place, becoming part of the ranching communities and defending these places through political activism. Sophie Sheppard, for example, organizes citizen meetings and petition drives and testifies at countless public hearings to protest the encroachment of military airspace and to halt the proposed construction of a geothermal plant in Surprise Valley. Each battle exacts enormous amounts of volunteer time and energy from the authors, who have full plates already. Defending place entails personal sacrifice, taking time away from families and vocations, and placing these women in the middle of conflicts. The women who lead these fights—who speak up and give time—gained the deep respect of the class, even those students uncomfortable with "bra burning" feminism, as one put it.

One Saturday, I led an optional field trip to Surprise Valley to visit two of the authors. Half a dozen hardy souls rendezvoused at seven a.m. in the parking lot of an Albertson's grocery store. We carpooled, each car armed with a walkie-talkie, and caravanned north, past Pyramid Lake, stopping near the Black Rock Desert at the off-the-grid Planet x Pottery. After topping several sets of ragged desert hills, we emerged into the astonishingly verdant Surprise Valley, where we had lunch with author Sophie Sheppard in the town's small café. At the adjacent table, we beheld a dozen elderly women, wearing bright purple dresses and red hats, laughing and talking boisterously, enjoying what appeared to be a regular gathering. Sophie showed us the spinning and weaving store that has become a community gathering place. She then took us to her house, where she served up homemade blackberry pie as we admired her many paintings. Before heading home, we visited the ranch of Linda Hussa. The rigors of ranch life were impressed upon us when we learned that she and her husband, in their big pickup truck, had to leave us quickly to retrieve a wheel that had rolled off their trailer, laden with cut wood. They were racing against the dark.

The trip was easily the highlight of the class, confirming the value of experiential learning. But whether they went on the field trip or not, many students singled out *Sharing Fencelines* as the most influential book of the semester. Their responses suggest that teaching the literature of one's own region engages students like no other. Nevada deserts have so relentlessly been depicted as barren, empty, and ugly—the perfect place to dump the nation's nuclear waste!—that students either share this negative judgment because they have been brainwashed by bad press or sheepishly hide their own fondness for the place. Books like *Sharing Fencelines* validate the love that some students have for the Great Basin and teach others to see it in a more positive light. Reading books about one's place has the effect of making a nowhere become a somewhere. Certainly regional literature should be part of a place-based course, but I'm now inclined to assign at least one local text in almost every course I teach.

Unfortunately, in our own time most landscapes are damaged, creating special challenges for teaching about place. But rather than ignoring real problems by reading only about good places, or becoming mired in despair by dwelling on bad ones (the environmental equivalent of the suicides and madness that darken much women's literature), we must expose students to narratives of restoration, to end with songs of hope. Interestingly, many writers tell a restoration story by creating a strong mother figure who takes maternal responsibility for her locale. Gloria Naylor's *Mama Day* may not at first appear to fit the theme of restoration, since Willow Springs, an island off the coast of the South Carolina–Georgia border, appears to be idyllic, far removed from the urban jungle of New York City, its foil. But restoration operates in two ways. Willow Springs was settled by ex-slaves and thus restores land ownership to blacks, who revive elements of African culture, suppressed during slavery. Too, restoration is metaphysical as the powerful forces of jealousy and hatred must be dethroned by "Mama" Day in order for her granddaughter's health to be restored. The powerful conjure-woman Mama Day became a focal point of class discussion as she recalls other older women in our readings, such as Mrs. Todd in *The Country of the Pointed Firs,* whose enormous importance to the community belies their actual physical size. Mama Day and her literary sisters *embody* place, being so integral to their places that it is difficult to imagine them living anywhere else or to think of the places without them. They are exemplars of the bioregionalist notion of self-in-place.

Ana Castillo's *So Far from God* likewise features a powerful maternal protagonist who takes care of her community. One day, fed up that her good-for-nothing husband, Domingo, has still not fixed the screen door, Sofi, mother of four girls, decides to launch a campaign for improvement.

Her domestic frustrations are projected onto the whole town of Tome, New Mexico, which, as Sofi observes, has a lot of problems that no one is doing anything about. So La Sofia runs for mayor, even though the town has never had nor missed one. Mayor Sofi persuades the people of Tome to establish a sheep-grazing wool-weaving cooperative, herding being a traditional land use threatened by wealthy gringo outsiders who are buying up the land. Miraculously, the economic viability of Tome is restored, people take pride in their positions of responsibility in the cooperative, and they are energized out of resignation.

By semester's end, my students had articulated a paradox that I had not foreseen. In positing the course, I was interested in the conflict between feminist values of liberation and self-assertion and bioregionalist values of community and commitment to place: self-centeredness versus self-centering, if you will. My students discerned a pattern in the literature whereby women who sacrificed for the greater good became empowered and stronger as individuals. One student, writing about Sophie Sheppard's essays in *Sharing Fencelines,* expressed the paradox this way:

> The self sacrifice that is needed for the life of a community and the required selfishness which exists in order for the life of an individual to prosper can easily be placed into a dichotomy. However, these two opposing sides do not have to be in a battle with one another. Ironically, in writing about the sacrifices that are necessary for the life of her family and community, the confidence in her own individuality is also emphasized. In fact, paradoxically, the community seems to make her as an individual stronger.

Another student, comparing *Sharing Fencelines* with *So Far from God,* observed:

What I find interesting and inspiring is that in both of these stories personal independence and communal respect are the rewards the women receive who sacrifice for the good of their communities. . . . As Sofi gives herself over to planning and organizing programs to help her community, she changes as a person. She is able to speak her mind to Domingo and Fe, and they in turn give her more respect than they ever had before. The lives of these women have been *enhanced* by the sacrifices they have made for their communities, on top of the community being helped as a whole.

The strong, community-minded women we met in the literature caused us to rethink feminism itself, honoring not just mold-breakers such as Sara Smolinsky but also women whose work is an extension of traditional roles. One student noted in an anonymous review:

I had been avoiding taking Women and Literature, assuming that the reading and assignments would emphasize and exalt "feministic values." From other courses I had taken at the university, "feminism" had been discussed primarily as women who break free from the submission of men and society; women who rebel against the traditional roles set up for them in society, such as being a wife and mother; women who do what they want, no matter what the cost for their families or their communities in which they live. In one sense it sounds wonderful: freedom. In another, it is selfish, and not the qualities that I admire in women that I know.

This class, however, helped me to come to a deeper understanding of the emotions that I have felt and struggled with over the last couple of years concerning these issues. In the literature we read, I saw strong, independent women fulfilling traditional women's roles.

The class thus increased students' awareness of and respect for women whose work is so often taken for granted. It may take a village to raise a child, but it also takes "mothering" to care for a village.

It is admittedly tricky for a women's studies course to advocate selflessness and service, for those very ideals have undergirded women's subjugation. However, just because those ideals have been invoked to keep women down does not mean that the ideals themselves are unworthy. To the contrary, we need to elevate their value. One thing this class accomplishes is to acknowledge the tremendous *bioregional* importance of what has historically been coded as women's work. Whereas feminism rightfully advocates giving women access to traditionally male-dominated spheres, one of the less fully realized potentials of the movement is to *inspire men* to make common cause with women in caring for families, community, and place, and to redefine those caretaking activities in non-gendered terms, as adults' work, the performance of which is a mark of maturity. My students—female and male—broadened their understanding of home and seemed genuinely interested in helping others, defending places, and strengthening the social fabric. I did not observe a marked difference in the responses of male and female students to the material. On the contrary, it was remarkable to witness the extent to which the men in the class strongly identified with the female protagonists of the works we read. For all the students to see their own community-spirited impulses honored in literature and in class seemed to confirm their intuition that there is a nobler, more meaningful path than looking out for number one, the message that bombards us in popular media, especially advertising. "One of the conclusions that I've reached," wrote one student, "is that a great part of what makes a home is sacrifice; sometimes, the greatest things in our lives are not without a cost—but it is a price that we are often willing to pay, despite feelings of reluctance, fear, or uncertainty. At times, it may seem as though we have much to lose; however, in retrospect, we find that we've actually gained far more than what we had lost."

NOTE

My sincere thanks to the students in English 427A, "Women and Literature," Fall 2005, for their dedication, good ideas, and permission to quote from their work.

COURSE READINGS

Castillo, Ana. *So Far from God.*
Chopin, Kate. *The Awakening.*
Hurston, Zora Neale. *Their Eyes Were Watching God.*
Hussa, Linda, and Sophie Sheppard, and Carolyn Dufurrena. *Sharing Fencelines.*
Jewett, Sarah Orne. *The Country of the Pointed Firs.*
Naylor, Gloria. *Mama Day.*
Yezierska, Anzia. *The Bread Givers.*

WORKS CITED

Beecher, Catharine. *A Treatise on Domestic Economy.* 1841. Reprinted in *Five Hundred Years: America in the World,* edited by Scott E. Casper and Richard O. Davies, 129–32. 4th ed. Boston: Pearson, 2005.

Castillo, Ana. *So Far from God.* New York: Plume, 1993.

Chopin, Kate. *The Awakening.* 1899. Reprinted in *The Awakening and Selected Stories of Kate Chopin.* New York: Pocket Books, 2004.

Friedan, Betty. *The Feminine Mystique.* New York: Norton, 1963.

Hurston, Zora Neale. *Their Eyes Were Watching God.* 1937. Reprint, New York: Perennial Classics, 1999.

Hussa, Linda, Sophie Sheppard, and Carolyn Dufurrena. *Sharing Fencelines: Three Friends Write from Nevada's Sagebrush Corner.* Salt Lake City: University of Utah Press, 2002.

Jewett, Sarah Orne. *The Country of the Pointed Firs.* 1896. Reprint, New York: Dover, 1994.

Le Guin, Ursula K. "The Carrier Bay Theory of Fiction." 1986. Reprinted in *Dancing at the Edge of the World: Thoughts on Words, Women, Places,* 165–70. New York: Grove Press, 1989.

Naylor, Gloria. *Mama Day.* New York: Vintage Books, 1988.

Snyder, Gary. "The Place, the Region, and the Commons." 1990. Reprinted in *The Gary Snyder Reader: Prose, Poetry, and Translations, 1952–1998,* 183–99. Washington, D.C.: Counterpoint, 1999.

Thelma and Louise. Dir. Ridley Scott. 1991. Metro-Goldwyn-Mayer, 1996.

Welter, Barbara. "The Cult of True Womanhood: 1820–1860." *American Quarterly* 18, no. 2 (Summer 1966): 151–74.

Yezierska, Anzia. *Bread Givers.* 1925. Reprint, New York: Persea Books, 1999.

CHAPTER 6

The Complexity of Places

SUEELLEN CAMPBELL

There's something compellingly immediate about places. Dirt, trees, houses, mosquitoes, a barking dog, the weight of the air: these things are so directly physical, so very *present*. Our personal reactions and associations draw us into their orbits as well: I caught tadpoles here as a child; I dream often of that house; this is where we watched those elk at twilight. Such sensory and emotional elements create our ordinary experiences of place. Then, too, we all see what we know about. Where English teachers see the settings of books by Thoreau or the Brontës, geomorphologists see the shapes of rivers and hills, historians see contested territories and changing land uses, ecologists see carbon budgets and energy circuits, painters see colors, shapes, and values, toxicologists see lingering poisons—and so on, in seemingly endless variations on perspective, specialized lenses, and kinds of vision.

Sometimes it's enough just to enjoy the feel of sunlight on skin or to think only about mineral uptake in a single grass species; some purposes require simplicity and focus. Yet these personal responses can make us feel that our understanding of any given place, any landscape, is complete, when of course it never is. For all places are endlessly complex—intricately composed not only of the immediate and personal but also of what other people can see, know, and remember; what is present but invisible; what is past and future, the deep time of geology and evolution, climate change, former human inhabitants; even what is somewhere else, like global economies, sources of acid rain, and migratory birds. Non-human and human elements are always woven together, too, an inevitability that makes "landscape" a useful word, combining as it does the non-human elements of land with our human scope of understandings, relationships, and effects.

This is the challenge that over the last decade has shaped much of my work as researcher, writer, and teacher: to think and learn about the complexities of places. And I do mean challenge. There's so much I don't know and don't really know how to learn, especially, though not only, in the sciences. It can take many frustrating hours to find even the simplest information, facts and concepts so basic to those in the field that books don't explain them, such as "Why *is* carbon so important?" or "What counts as a salt?" And it's hard to decipher and translate the language of unfamiliar disciplines, language that can sound deceptively familiar: a sentence like "What is needed is the incorporation of an eruption's SO_2 production into its magnitude" looks easy enough but makes little sense without the right background and context. Still, I believe, understanding more fully where we live is important enough, and rewarding enough, to be worth the trouble.

What I want to do here is look at what this task can mean in practical terms. First, I'll give you an example from the book I wrote that started me along this path. Then I'll talk briefly about the book I'm currently writing, one that foregrounds the multiple ways of understanding landscapes. Finally, I'll lay out a teaching exercise that aims to coax others into undertaking similar explorations.

I'll start with some lines from the beginning of my book *Even Mountains Vanish: Searching for Solace in an Age of Extinction* (University of Utah Press, 2003).

> We had driven in earlier that afternoon across the Pajarito Plateau, past the bold signs for Los Alamos and into Bandelier National Monument, where we'd stopped at the rim of Frijoles Canyon to have our sandwiches in the sun. . . . I found a smooth spot on the ground, rolled my jacket into a pillow, and stretched out on my back. The air felt cool and clean, the sun warm against my skin, my every breath was fragrant with piñon and juniper, and a pair of jays above me flashed like blue coal against the sky. I lay still, tried to stop the clatter of voices in my head, and waited for that infrequent but familiar feeling that some space inside me was opening up, making room for a kind of peace. (1–2)

On that February afternoon in 1996, I knew only a little about this place, just what I'd noticed on a handful of previous visits and a few things I'd picked up in some recent reading. What I found that day wasn't the peace I'd hoped for but a set of questions that over the following years kept expanding and deepening, leading me into months of intensive research and the slow writing of a book about my inquiries. Today, if I were to revisit that opening scene, I could add these kinds of details:

> I lie on my back in Bandelier National Monument in northern New Mexico, surrounded by the Sangre de Cristo range of the Rocky

Mountains. My feet point east across the Rio Grande (both river and 35-million-year-old rift) toward the Great Plains; the top of my head points west toward the Jemez caldera. Beneath me is the thick, million-year-old, volcanic Pajarito Plateau, and below that lie the compressed traces of billions of years of continental drift, changing climates, advancing and retreating oceans, species evolving and then going extinct (including the enormous Seismosaurus, perhaps the largest land animal ever), sometimes slowly, sometimes in sudden cataclysms. (There's an especially visible instance not far from here of the K-T or Cretaceous-Tertiary Boundary, a layer of ground rich in iridium that could only have come from an asteroid collision, a layer that marks the end of the dinosaurs).

On my left is Los Alamos National Laboratory, home of the first nuclear bombs, the weapons that brought the end of World War II (and thus, my father says, the possibility of my own existence, since he returned alive from Iwo Jima) and worrisome amounts of radioactive waste with very long half-lives, some of them many thousands of years long, one (U-238) as long as this solar system is old—and home, as well, to sites sacred to the people of the nearby pueblos. On my right is Frijoles Canyon, where some eight centuries ago the ancestors of these people built their houses high in the soft volcanic tuff of its golden cliffs. The ruins of their kivas and other ceremonial and communal buildings lie along the canyon floor among tall, vanilla-scented ponderosas (five hundred years old), curly-leafed Gambel oaks, mule deer, coyotes, Abert's squirrels, and short, sturdy green horsetails (one of the oldest land plants, some four-hundred million years old, whose forebears grew as tall as trees).

I could easily go on, given everything I learned in several months of (mostly) library research into geological and evolutionary history, the development of the atom bomb, and other elements that I came to see as core parts of this place. How I learned to do this research became part of the book as well. At first I was comically inept, but I improved, thanks to many, many hours of floundering and a few key "methods": most importantly, patience with my own ignorance; frequent advice from librarians; trial and error on the computer, where it matters hugely which search terms you use, in which databases; browsing in the stacks, which is usually the best way to find the right books; and the discovery that recent college and graduate-student textbooks can cover in a reasonably understandable way a lot of the ground a non-specialist needs. For the book's later chapters, which involved three other places, I also took field seminars at a nearby national park, interviewed scientists on my campus, talked and listened to the people who lived and worked in the places I traveled to visit, and paid a slightly different kind of attention to my own daily life and home.

During all this time, I thought about how many things are important to specific places that aren't visible, at least when and how we're looking—things like radioactive wastes, the fats in a marmot's summer diet that determine whether it will survive its winter hibernation, the effects of vanished ice ages, distant wars, plant roots, the cultural traditions of human beings, and so much more. I read voraciously, noticed details, clipped newspaper pieces, and accumulated boxes full of paper and books. Not least, I kept asking myself questions and trying to make connections. As I wrote, learned, rewrote, and learned more, I came to realize that I truly was developing richer ways of seeing and understanding.

A GUIDE TO LANDSCAPES

Once my appetite for such complex matters had been whetted, I embarked on an even larger project, a guide to landscapes in nature and culture. This time I arranged for help, mainly by enticing English ecocritic Richard Kerridge to be my cowriter, but also by recruiting as advisors and sometime contributors Australian ecocritics Kate Rigby and Mark Tredinnick, several scientists and an environmental historian from my university, and other friends and colleagues. The book has several components, but so far most of my energy has been focused on learning about large landscape types like wetlands, deserts, and tectonic /volcanic landscapes.

Generally speaking, my procedure is to gather information from as many angles as I can and then try to figure out some of the intricate interactions among natural and cultural actors and effects. As an English teacher, of course, I think about literature: which writers have been most key to this kind of landscape, and how; what kind of "character" has this landscape typically been in literature; what kinds of meanings has it suggested to our culture and to others, and why? I think about words, too: how their fuzzy edges and overlaps complicate the fuzzy edges and overlaps in the physical world, and how in their etymologies, changing usages, variations, and distribution they mark cultural and political history as well as local and specialized knowledge. But I'm also working to fill in my blind spots, some of which are blinder than others. Supposing I were an art or film historian, I ask myself in a relatively easy instance, what paintings or movies best show the ways this landscape has been represented or how it has affected human affairs? Or the tougher questions: What would a climate historian say or ask about this landscape? An environmental historian? A plant ecologist or wildlife biologist? A geologist? A sociologist? I often ask for help (and am always perversely relieved when the answer to my question turns out to be complicated rather than simple: no wonder I couldn't figure it out myself), but mainly I collect tall piles of books from all over the university library and try to read them myself.

I've been finding that I oscillate between two types of questions. Many are broad and basic: What geological and climate forces typically create this kind of landscape? What kinds of human economies have been tied to these places, and why? At a sufficient level of generality, some of these matters are widely shared: deserts do share some characteristics, as do volcanoes, grasslands, mountains, and in some ways, the human cultures associated with these places. Other questions are specific, even idiosyncratic, and change from one landscape type to the next: Why has indigo been an important dye in the Sahara and Arabian deserts? or Where exactly was *Lawrence of Arabia* filmed? or even When might that petroglyph have been created, and what might its creator have eaten for supper?

Of course we can't hope to cover everything in this book, so we've set ourselves some restrictions (we're focusing mainly on the English-speaking world, for instance), reconciled ourselves to being far from comprehensive or encyclopedic, and given ourselves permission to pursue the questions that arise from our own interests and experiences and also from what we learn as we research. The resulting chapters have their own structures. We've had to divide part of our wetland material according to location, since English and North American categories truly don't match, in either cultural or scientific terms: England's "swamps" are America's "marshes"; the "fens" in England are drained, while Canada's are not; English scientists think in terms of mires, moorlands, and peat, while Americans often, or usually, do not.

By contrast, the chapter we're calling "Landscapes of Ice" currently includes sections named "What and Where Are Glaciers?" "The Last Glacial Maximum and Since," "Ice and Today's Landscape," "Ice and *Homo sapiens*," and "The Little Ice Age, Glaciology, and the Sublime." Along with explanations of the formation and behaviors of glaciers, "recent" ice ages, what ice core data from Greenland say about global warming, and other scientific matters, this chapter touches on such "cultural" elements as how Walden Pond is in a glacial kettle, which accounts for the depth, clarity, and self-containment that impressed Thoreau; key links between climate and human development, dispersal, and cultural stability; how priests tried to stop glacial surges with prayers and exorcisms; the setting of Mary Shelley's *Frankenstein;* John Muir's story of the dog Stickeen; and, as advertisers say, much, much more.

The challenge of assembling, deciphering, organizing, interpreting, and integrating this variety of information continues, indeed increases, the more we work on this project, and it has been a constant struggle to avoid feeling too overwhelmed to proceed. On the other hand, the satisfaction we are taking in learning—and making sense of what we learn—continues to increase as well.

THE TEACHING EXERCISE

Not surprisingly, all this work has spilled over into my teaching, where I've been looking for ways to lure students and colleagues into thinking about places from similarly multiple and complex perspectives. One of the tools I've developed is the exercise that follows, along with some comments about how I've used it and to what effects. This is a fairly generic and compressed version, one that can be done quickly or slowly but might ideally last either about seventy-five minutes or a whole semester. (For a longer version of the exercise, see *ISLE* 13(2).)

Today I'm going to ask you to think about how layered and complicated places are—and how many different ways we have to perceive, think about, and understand them. This exercise is divided into stages, and I'd like you to try to think about just one thing at a time.

Choose one place that you know well, maybe a place you love. Put yourself there in your imagination and memory. Choose your season, time of day, and weather. Now settle in: sit down in a particular spot; close your eyes, take a deep breath, quiet your mind.

In your imagination, open your eyes. Look around you. Choose just one part of what you see to focus on, something close enough to you to fit into this room, maybe close enough to touch. This will be the "place" you'll be thinking about for the next forty-five minutes or so.

1. What do you actually see, with your eyes, right now? Forget what you know and think only about what you see. Be concrete, detailed, and straightforward—the visual facts, but precise. Avoid metaphors. If you don't know or can't remember something, think instead about questions you could investigate. Think about lines, shapes, sizes, balance, design. Think about textures. Think about colors—their shades, intensity, variations, how they reverberate against or meld into each other. Think about what painters call "values"—if you were to shift into black-and-white vision, what would be lightest, what darkest, what in the middle? Stay concrete and detailed. Think about light and shadow, how they change colors, how they move.

Imagine that you have a few other lenses at hand: zoom in close, use binoculars, a magnifying glass, a microscope. Or zoom back, use a wide-angle lens or a telescope. Look up. Look down. Hang your head upside down. Take your glasses off. Put on rose-colored sunglasses. Squint.

Think about how all these visual parts go together.

2. Pay attention to the rest of your body, all the other ways you can sense this place. Touch, textures, temperatures? Tastes? Smells? Sounds?

3. I asked you to choose a place that you know well. How and why do you know it? How do you feel about it? What memories does it hold for you? Emotions? Are your memories and emotions deeply personal and idiosyncratic, or are they things you'd expect other people to share? Do you think that your own identity, or your sense of yourself, the shape of your life, how you matter to yourself, is somehow tied up with the identity of this place?

4. Has anyone made art or written a poem, story, essay, or book about this place? What human events have happened here? Who has lived here, and how? How has this place been tied to events happening elsewhere, through commerce or politics? Who owns it? Who controls what happens to it? How have different parts of our culture thought about this place, or other places much like it? What have humans done to hurt it, or to help it? Pollution, poverty, warfare, invasive species, habitat loss, climate change, strip mining, deforestation, desertification, hurricanes, golf courses or ski areas or housing developments, disease? What forces are working to fight these threats? What lines of connection, influence, or effect could you trace going out from here to other places, or coming in from them?

5. What is sharing this place with you, right now? People, plants, other animals, birds, insects: what kinds of lives do all these creatures live here? What's the temperature? What's the air like? What's beneath

you? What kind of rock, or soil, or plant cover? What is here that you can't see—because it's underground or inside something else, because it's so small, because our human senses can't respond to it? How do all these elements of this scene interact with each other?

6. What happens here when you're not around? How long has this place been the way it is now? What was it like a hundred years ago? Twenty thousand years ago, when average global temperatures were roughly ten degrees Fahrenheit lower than they are now? Five million years ago? A hundred million? What might it be like fifty or a hundred years from now? What might happen as the planet continues warming, say another four or six degrees Fahrenheit, as some fairly conservative models predict, making it warmer than it's been since humans evolved?

7. Does this place *matter?* Think cosmically, philosophically, spiritually. What would be lost if it were destroyed tomorrow? Might this place have anything to teach us, any kind of wisdom we might want? What happens to the place itself when we start asking such giant questions?

Finally, think about what happens when you do this exercise. Which parts were the easiest, which the hardest, and why? Where do you have the most questions, the fewest facts or ideas? What happens in your head when you do these as a series? How do these ways of thinking affect each other? What would change if you focused on a different place? Why did you choose the one you did?

This exercise began its evolution in my senior-level nature writing course, a class that's equal parts literature course and writing workshop. Typically fewer than half the students are English majors, many of whom read and write pretty well but know little about nature; the others, who come from all over campus, often know a good bit of science but not so much about reading or writing.

One fall I spent the first two class sessions outdoors leading an exercise I called "Eleven Ways of Looking at a Tree." It worked best as a diagnostic tool, showing me how hard it was for those students to focus, how quickly they abandoned concrete observation for vague abstraction, how much their initial reactions limited them. (I suppose I shouldn't have been surprised by this, but I was.) Many responded to my opening instruction to write down only what their eyes actually saw by describing how big trees made them feel peaceful and reminded them of trees they loved to climb as children. And many seemed uninterested in thinking about much else.

A couple of years later, I tried the same "Eleven Ways" much later in the term, when my students had spent more than two months immersed in the genre and being pressed by me to be concrete and specific. I used only one class day outside (Election Day, when concentration was difficult), sitting under a single cottonwood, but suggested that those who wished could transform the highlights of their notes into short essays. The results were much better, with many students telling me they'd found the day stimulating and would try to do the same thing as they worked on their next essay.

That same fall, an ecologist I know had invited me to speak to faculty and graduate students in the College of Natural Resources at my university. I was to be part of a lunch-hour series including an environmental historian, an ethicist, a political scientist, and others. "Talk to us about the one thing you believe it's most important for us to think about when we think about the land," he had said. Well, I decided, putting things together into a focused but complex picture is one of the things I often try to do, and I do believe it is important. Fifty minutes, though, is not very long. So I adapted my exercise, abandoned the writing element, combined my eleven ways into seven, added some harder questions I figured my listeners should be able to answer, and asked these men and women to imagine themselves in a place they knew well and loved. Here, as I had done before, I began with the most concrete and immediate questions, then moved through those I thought they'd be most accustomed to thinking about to those outside their professional realm, and ended with the most cosmic. (I turned off the lights, spoke in a slightly incantatory way, budgeted my time in advance and kept an eye on my watch, spaced out the sub-questions, and watched for about half the group to open their eyes or stop jotting notes before I moved on.) The following summer, in an even speedier version I offered at a session on "Imagining Invisible Landscapes" at the 2005 meeting of ASLE, I asked the cultural questions before the science questions for the same reason; this is the version I've compressed above.

Other versions of this exercise are easy to imagine. Indoors or outside, involving writing or not, lasting an hour, a day, a whole term, for beginners or professionals, scientists or artists or humanists, focused on the place where the exercise is occurring or on somewhere else— there are many possible variables. The order of questions should be audience-specific; for beginning students, for instance, it might be best to open with questions about feelings, then work toward more-concrete observations and harder speculations. The wording of the initial charge ("Choose a place that _____") is also variable and critical: choose a place that you fear, think of the place you spent most time as a child, think about where you live now—such different openers would produce quite different responses. Of course there are many other questions to ask, with many different possible ways to focus.

What follows the exercise will also vary by circumstances. One might have the students type up their notes and turn them in, choose high points and craft a short essay, or use their unanswered questions to direct their research for a larger project. A good conversation can also ensue. As one ASLE listener said, debriefing might be important; he had chosen to think about a seriously threatened place he couldn't find a way to help protect and so had found the exercise very painful. A graduate student in ecology told me she, too, had found it painful: she'd realized that not only could she not answer most of the questions about the place she'd chosen, but worse, she couldn't answer them about any place. A senior wildlife biologist, on the other hand, remarked that although he'd thought about all those things separately, he had never put them together. The final paragraph of the exercise above might serve as a template for a follow-up conversation.

I've found this and similar sets of questions helpful in teaching nature and environmental literature as well. Few books attempt to be as comprehensive as the exercise might encourage (nor do they need to be), and so it's useful to think about which kinds of issues a book does address, which ones it does not, and what implications and ramifications ensue. Barry Lopez's *Arctic Dreams* is a rare instance of a book that considers most of these questions, as do Edward Abbey's *Desert Solitaire* and Robert Sullivan's *Meadowlands,* yet these books differ considerably in *how* they ask and answer them. Rebecca Solnit's *Savage Dreams* asks some of them very carefully; Annie Dillard's *For the Time Being* brings all its power to bear on even fewer. The books I read for other classes can sometimes be illuminated by this measure, too. Tim O'Brien's Vietnamese jungle, the Joad family's dusty plains and fruited valleys, Clarissa Dalloway's London streets, and Jane Eyre's tumultuous moors are all places important to their human stories, though the ways these writers consider place, the questions they address, are not the same.

Finally, I want to note that these questions will lead to quite varied results even when several people ask them about the same place or one person asks them about several places. Always, though, they are simple but powerful tools that can help us coax or push ourselves beyond our ordinary personal responses to a place, help us get past ourselves. And as I have said, the rewards are considerable: sharpened perception and understanding, richer insights to feed our writing and teaching, more of the knowledge and complex vision we need to act wisely and effectively as we work to protect what we value of life on this planet.

CHAPTER 7

A Place at the Table
Writing for Environmental Studies

JEFFREY MATHES McCARTHY

> "And how does cooking a meal teach college students to write?"
> —Interview question from a *Salt Lake Tribune* reporter

I'm not the nervous type, but when the second TV station called, I began to feel a little queasy. Three radio interviews. Reporters from the two major newspapers. And suddenly TV cameras were coming to class. The dean was coming too, the president left a message, other faculty were going to be there, and fourteen freshman composition students would be the hosts.

These freshmen were the center of attention for their place-based class project—a full Thanksgiving dinner of local ingredients. But why were they making such a meal and how did this project fit the goals of a writing class? Few professors would be inclined to describe their students as "diet aware." Detachment from our meals and from our land is as American as apple pie, especially for those accustomed to cafeteria food and raised in the suburbs. But when these composition students read about American eating, they found that the topic led beyond their plates to the towns and streets where they grew up, and from there to issues of global environmental health.

A week before Thanksgiving they were primed to serve up food, followed by their reasoned claims about the environmental consequences of the American diet. The regional meal would incarnate their research about our food and present an alternative for anyone willing to listen and chew. From my office it looked as if this project was going to be either a moon shot of a success or a *Hindenburg* of a disaster. I took a deep breath.

These fourteen students made up Westminster College's first environmental studies "learning community," which meant that they were enrolled simultaneously in a composition course and an environmental biology course. The reading they did in my class informed the field trips they took in environmental biology, and the work they did there contributed to the writing projects they undertook in composition. Such learning communities give first-year students an established peer group, energize interdisciplinary education, and promote retention. This local meal project was drawing attention because it put theory into practice by linking student action with environmental writing.

When the *Tribune* reporter asked me, "How does cooking a meal teach college students to write?" I said that the best writing mixes direct experience with careful reading. I added that students who take responsibility for something generally write better about it because they're moved from abstractions and generalizations to the particular qualities of the idea—or the beet they're preparing with their own stained fingers. Since I teach composition with a rhetorical emphasis, we study useful structures for argumentation, yet leave the particular shape of the final written argument to each student's interest and ability. As I told the reporter, students have to care about what they're writing for this rhetorical approach to work.

Though we don't wrestle too much Greek in my class, my students read

John Gage's *Shape of Reason* and apply its terms—thesis, enthymeme, dialectic—less as inheritors of classical rhetoric than in the tradition of American pragmatists aiming to do work with language. Looking back, I see that the regional meal was their direct action, and from that action came a feeling for writing's purpose and power.

The semester itself was organized into three main themes: Wilderness, Suburb, and Action. Those may seem curious at first blush—suburb?— but they emerged from my determination to engage students with the environments they actually inhabit. Because the idea of wilderness is so central to American self-definition and the Utah experience, readings from Roderick Nash, the Old Testament, Robert Kennedy Jr., and Chief Seattle went over easily. I complicated the journey with readings from Joyce Carol Oates and William Cronon. And thanks to the contributions of my colleague in biology, Ty Harrison, students studied our valley by canoeing its rivers, prodding its plants, and exploring the old Lake Bonneville shoreline.

The second theme, suburb, may seem surprising for an environmental studies program. But once we recall where college students come from, where many return for holidays, and where mainstream American culture resides, studying the suburbs as a place, a threat, and an ideal makes sense. Our readings came from sociologists, Web chatter about "soccer moms," and nature essays by Annie Dillard and Rachel Carson.

The third section of our course was action. Here we showed that the response is as important as the investigation and studied some of the ways people have reacted to environmental and political crises. We read Dave Foreman and Edward Abbey on direct action, Thoreau on civil disobedience, Wendell Berry and Bill McKibben on food, and Jon Waterman on lobbying Washington. In sum, the students looked at grand environmental themes, considered those themes in the local context of the Salt Lake Valley, and thought about how people who care can respond to environmental degradation.

The microphones and crockery and cameras made the students anxious, but I reminded them that things were actually fairly simple after all. The assignment was to prepare a meal of locally grown food, every bite planted, packed, and purchased within the magic ring of one hundred miles. That's all. The students chose to cook a full Thanksgiving dinner, we decided to tell the college press officer, and there we were: drumstick-deep in reporters.[1]

"But why should the radio stations care?" came the sharp question from a student who'd been more often on the dull side that semester. "I mean, if we were spiking trees, sure, but why all this attention for food?" During the next class we batted this question around and decided that our project had the "perfect storm" factor— three mutually intensifying forces that brought us the bright lights. The foremost media draw was simply seasonal: they were looking for Thanksgiving stories, and we served them one on a platter, literally. More substantively, there are people in any community—including Salt Lake—with a deep regional awareness, and a corresponding curiosity about efforts that prize local products. And finally, my students believe that Americans are ready to hear about the benefits of local agriculture and the consequences of agribusiness. It doesn't matter whether Utahns read Wendell Berry and Jane Brox and Gary Nabhan, or shop at Wild Oats, or wonder why their lettuce comes in a truck all the way from California. They are poised to be convinced, influenced by good ideas about food presented in clear language. The Wasatch Front media care because in November they need stories with turkeys, but more precisely, our project grabbed a "question at issue": the debate about modern food.

In an environmental studies learning community, it's important that we understand Kenneth Burke's assertion "rhetoric is the manipulation of men's beliefs for political ends" (41). Students arrive in Freshman Comp thinking of writing as a college necessity they'd best polish or as something they've always kind of fancied in a slightly embarrassed way. In this writing course, I try to direct them toward specific points of disagreement, where the written word can lead to action. Here students can motivate themselves by becoming conscious of that work's significance and writing's power. Gerard Hauser defines rhetoric in terms of its influence:

> Rhetoric is an *instrumental* use of language. One person engages another person in an exchange of symbols to accomplish some goal. It is not communication for communication's sake. Rhetoric is communication that attempts to coordinate social action. For this reason, rhetorical communication is explicitly pragmatic. Its goal is to influence human choices on specific matters that require immediate attention. (8)

Cooking this meal, talking to the press, and thinking about ecology made these students live their writing as an attempt "to coordinate social action." Also, this project helped students place the authors from our reader, *Saving Place,* in a rhetorical tradition. Indeed, I asked them to what extent the beets and potatoes now on the news were like language, symbols applied "in an exchange of symbols to accomplish some goal," and two of them said the potatoes were "our spokesmen, but better." Ultimately, theories of rhetoric were supplanted by practice, and action was the motivation for the students' getting messy with berries or smoking a trout to feed to classmates.

Strong and convincing writing was this group's response to the meal's insistently personal and direct experiences, and the primary assignment was a five-page argumentative essay. Their cooking and gardening and questioning led the students to generate assertions and questions that mattered to them and to local people: Should the college buy food at the farmers' market? Can Salt Lake families be healthier eating only seasonal foods? Should western cities preserve farmland from suburban sprawl? The point here is that the students nurtured their own ideas into powerful expressions of meaningful evidence. The payoff in composition courses is exactly this—a student finds the tools to build notions into ideas, ideas into assertions, assertions into convincing analyses. But I can't just give them assertions, or hand out pre-formed theses like Ready to Bake!™ Pillsbury dough. No, the students must mix and flavor and bake their own personal ideas and in the process find nourishment.

A few months' perspective helps me see that this latter dynamic—the move from the abstract to the personal in writing—is the same move environmentalists hope people will make in cherishing local food and weather, plants and streams. But did every one of these students cook themselves into better writing? Well, yes and no. I wish I could say "absolutely!" and show that their writing surged forward—but that's not true. Some writers struggled with the usual organizational issues, and others were vexed by the familiar demons of procrastination. This project wasn't a panacea, but it did succeed in leading students to topics they believed in. One student explained, "I cared about this one and wanted more time to do it," while another insisted, "This matters to me, and I'm going to show the people I interviewed what I wrote."

Cooking a meal hasn't turned these college students into Montaigne or Dorothy Parker just yet, but it made them care, and that's an important first step.

There were plenty of things for students to learn from this process: the power of direct engagement; a sense of group responsibility; how political process and public discourse overlap; how the everyday life of suburbs and food can become the stuff of sustained inquiry; and writing from direct knowledge of specific efforts.

From my end, the intellectual issues were clear and energizing for my life as a member of the college, and as part of this civic community. For instance, any trip to the cafeteria raised issues that Wendell Berry defines as "the politics of food." The same jaunt through the student union can illuminate the obesity and diabetes epidemic emerging from a diet of convenience. Our viewing of the movie *Supersize Me* connected all this to the corporate profit culture. The most interesting aspect of this place-based food project was a burgeoning regional awareness that prompted students to shape an ecological self in relation to their bioregion. There's ample fodder for argumentative essays here.

The focus on food empowered teaching because it revolutionized students' understanding of this place. It also enacted the interdisciplinary thought that we preach, identified by John Muir: "When we try to pick out anything by itself, we find it hitched to everything else in the universe." Matt said to his classmates, "I didn't care much about food or writing, but now I can't stop writing about food." Yes, I thought, the extra work will be well worth it if these freshmen read, write, and cook themselves into a conscious relation with the place they live.

Experiential education encourages students to surprise me with abilities I never see in the conventional ebb and flow of read-discuss-write, read-discuss-write. Of course, the meal was under the microscope from the local press and powerful administrators, so I was a little gun-shy about big surprises. All this "empowering" and "revolutionizing" carries risks, so scripting the process with a particular setting and a particular assignment was crucial. Finding an inspiring place for our meal, beyond the classroom's metal desks and away from windowless conference rooms was the first step. I reserved a light and airy alcove in the theater building and had campus catering set up tables with cloths and silverware and glasses and water in pitchers. In those shiny circumstances we were already having a good time.

Asking students to write two short pieces that week also shaped the conversation. The first was a paragraph response to Bill McKibben's line "In the end, real environmental progress may depend on remembering that we live in particular local places, not just in a globalized world" (55). The second, due at the meal, was a simple description of what food each of them brought, where they found it, and how. These brief writings had the students thinking about just what intellectual, academic, political, and cultural issues were at stake in every bite of potatoes and every sip of grape juice. Moreover, these writings positioned them to shine when the provost asked questions or a reporter wondered why this was making them better writers. Within the field of environmental studies generally, it's important to show that a project is more than a gimmick or an excuse to get dirty. For these students, the dividend of excitement was reinvested in their writing.

As much as I tried to script the process, the best learning moments came from the free interaction of surprising ideas with unexpected influences. For instance, the turkey showed up pale, slumped, and shriveled. It looked bad, and the students looked worried. Here we were, promising Norman Rockwell's own feast, and the bird looked like it had a Mardi Gras hangover. We had ten minutes before the guests came, and the TV crews were already setting up. Enter the college chef, our knight with shining cutlery. His white coat highlighted blue-tattooed forearms and the cigarette behind his ear. He won our smiles when he told us about the satisfying consistency of organic birds, the false plumpness of store-bought turkeys, and the overall evil of thanksgiving meals pumped with chemicals. I stepped back, happy to see him slicing away at the bird and at our preconceptions. What had seemed likely to demoralize our little group instead gathered them in a circle of nodding heads and relieved chuckles.

Another surprising moment came with the bright TV lights. As soon as the cameraman wanted an interview, the students scattered like pigs from a gun. Except for one. Mia had been a quiet contributor all semester—more thoughtful than boisterous. Suddenly, all the loud and rowdy ones were hiding behind each other while Mia strolled confidently to the counter and gave a composed and calm message that would have put Barbara Walters to shame. With her blond head tilted to one side, she recounted her own adventures at the chain superstores and then tapped one finger to punctuate her summary of Wendell Berry's alternative. It is always gratifying when a class exercise coaxes out surprising skills from students and positions them to feel more capable and accomplished than they did the night before. This project reinforced my determination to give college students authority, and live with the consequences.

The *Deseret News* reporter focused her story on one student's new familiarity with local organic farmers. Madeline described the farmers she visited: "They weren't Amish, they drove cars, the kids go to swimming lessons—but the farm gives food to the whole family, and makes money too. Growing what you eat is a healthy way to live in your place." When Madeline developed a relationship with local food producers, she learned that her local environment is both powerful and vulnerable. The purple potatoes Madeline brought us were a mashed favorite, and for her became a symbol of place. Eating local food encourages a conscious relation to the local environment, arguably the best starting point when teaching environmental studies.

There were several other surprising learning moments in the hubbub of the meal. While we passed each other locally grown and pressed grape juice, discussed the merits of our apple vinegar salad dressing, and enjoyed a course of smoked trout, the students were center stage for various audiences. People walking past saw four tables, and at each of them a freshman explained just how a certain vegetable relates to suburban sprawl, or why it was better to get a given potato from our own valley than from Washington State. Finally, my students noticed that in radio interviews they had to be absolutely concise and direct. Jake and Alex later told me that it was like all our work on thesis statements was magically reincarnated, because to convince a reporter the main idea needs to be directly expressed and then followed with a vivid illustration. I beamed to find them putting into practice abstractions like thesis, topic sentence, and evidence that we'd wrestled for two months.

Did the students mind the extra work? Inspiration attends responsibility, and the day's excitement multiplied it all into gustatory and intellectual pleasure. I heard Meg tell a reporter, "It wasn't as hard as I thought to find the ingredients, and it tastes a lot better than the usual food." They were proud of their work, and proud of the thinking that they shared with the invited guests.

We all know that taking classes beyond the classroom doesn't always go smoothly. I resent the in loco parentis dynamic of extended field trips, and I'm impatient with slow walkers on hikes. Just taking a group outside has its problems. This fine climate pulls back the curtain on my own preference for trees to chalkboards and leaves us all distracted and loud-voiced—as unsettled as our papers blowing downwind. Classwork that leads us physically into our places needs to walk a fine line between the satisfactions of getting outdoors to feel this world and the tempting distractions that accompany such junkets. A project that insists on student responsibility focuses the group and transcends distraction when it gives students a stake in their own learning.

Every composition class will have a few people who slide around the assignment as they would ski past the challenging line, who will go through the process with a smile but for a million reasons during that week in that month will be distracted. Still, the regional food effort made a big impression on most of these students, who talked to produce clerks, visited farmers, made assertions on the radio, and reported in the newspaper. The place-based learning community that united environmental biology with composition energized Westminster freshmen because it drew on the landscapes they knew, rivers they'd floated, fields they'd visited. The Salt Lake Valley became the center of their attention. When an essay started, "You could smell the dirt on the potatoes," I knew these freshmen were writing to explain relations they felt intimately and cared about personally.

But there's something else in play here, and it has to do with the academy's traditional embrace of theory and suspicion of practice. As I consider this regional meal assignment, it seems that the biggest payoff is bringing academic attention to this place we live. My experience of education is that those not living in New York or London often find their place slighted by abstractions that insist all that matters comes from a cosmopolitan city or, worse, some holy realm of ideas. It's a long way from smelling potatoes. The worst part is that this approach is endemic to literary studies, resulting in a denigration of "regional" authors, while our colleagues in the sciences and business are far more likely to choose local samples for laboratory and casework. In contrast to the way I learned and have for a long time taught, place-centered thinking fertilizes an ecological self, a grounded way of knowing that thrives in lived awareness of the places we inhabit and the characteristics that define it as a bioregion.

It's true I was nervous about putting my students in the spotlight, but I didn't need to be. The same pressure I feared would upset their rhythm motivated them into deeper thinking and stronger cooperation. If an environmental studies curriculum is going to succeed with students, it is going to generate action and position those students to see their world as a system they should engage and shape, instead of a given category to be memorized and accepted. Thoreau says, "Thought is nothing without enthusiasm," and the place-based meal project generates student enthusiasm in spades for the particular land they inhabit. It fosters confidence in their ability to live in concert with that land and recognizes them for acting on ideas they care about. Presenting students to the place where they live invigorated our writing classroom and gestured beyond to the society that these students aim to reseed and cultivate anew.

NOTE

1. Though every bit a Thanksgiving dinner, some of the difficulties—like salad dressing without oil, or dessert without cane sugar—became opportunities for surprising alternatives. We used apple vinegar for the salad, and baked apples in honey for dessert. One index of success is that the camera guys stayed for second helpings.

Here's the menu:

- turkey: Moroni, Utah
- potatoes, squash, and rosemary: Draper, Utah
- jellied berries: Westminster College campus, from the plants used in landscaping
- bread: grain grown and milled in Lehi, Utah, and baked with all-local ingredients in Logan, Utah
- grape juice: grapes grown and squeezed in Sandy, Utah
- trout: Strawberry Reservoir, Utah
- apples: Payson, Utah
- honey: from quiet bees at the Trappist Monastery in Huntsville, Utah
- butter: West Valley City, Utah
- Feta cheese: Tooele, Utah
- salad greens: Ogden, Utah
- beets: Santaquin, Utah

COURSE READINGS

Abbey, Edward. "Eco-Defense."

Berry, Wendell. "The Pleasures of Eating."

Burke, Kenneth. *A Rhetoric of Motives.*

Cronon, William. "The Trouble with Wilderness."

Dobrin, Sidney I. *Saving Place: An Ecocomposition Reader.*

Gage, John. *The Shape of Reason.*

Hauser, Gerard. *Introduction to Rhetorical Theory.*

McKibben, Bill. "Hunger."

Nabhan, Gary. *Coming Home to Eat.*

Oates, Joyce Carol. "Against Nature."

Thoreau, Henry David. *The Writings of Henry David Thoreau.*

CHAPTER 8

Meet the Creek

ELLEN GOLDEY AND JOHN LANE

The yellow public school buses will pull through the gate at White's Mill at 9 A.M. and disgorge sixty fifth graders from Chapman Elementary School. It's been raining for days, but finally it's clear. Our four learning stations are almost set up for "Meet the Creek," and sixteen freshman college students are busy spreading green and gray tarps on the ground and finalizing their plans to introduce four pods of wild kids to our learning landscape, Lawson's Fork Creek.

Lawson's Fork is a tributary of the Pacolet River and one of the headwater streams in the Broad River Basin. Here at White's Mill in Spartanburg, South Carolina, the stream is thirty feet wide and two feet deep at its deepest point, though the water level is up a bit today. The sound of the water falling over the dam just upstream is constant and soothing. The old dam, where a gristmill once ground meal for flour, was built from stone cut from the shoals now invisible beneath the pond. This old mill site is one of the most scenic spots on the thirty-mile creek, the only significant stream contained entirely within the political boundaries of Spartanburg County. Few citizens know of the stream's existence, let alone understand its role in the local ecology, perceiving it as merely a drainage ditch for urban runoff. Through "Meet the Creek" we introduce a new generation to this place we've come to love, a living stream on the path to recovery from an abusive past.

This stream, like the others in the South Carolina piedmont, has felt the strain of "progress." Nearly all the rivers and creeks in the region are classified as impaired by the South Carolina Department of Health and Environmental Control (DHEC). All are unsafe to drink, and most aren't fit for swimming. A few still can't support aquatic life three decades after the Clean Water Act took effect.

The floodplain downstream looks the way an outdoor classroom should look. We're surrounded by big tulip poplars and sycamores, and we know the children will enjoy digging their hands in the tons of sand that were deposited across the plain last week by high water. Where the old mill stood there's a rustic clubhouse used by the affluent subdivision now surrounding the property. Every year the homeowners are kind enough to let us bring the kids in for this day of discovery and excitement. We know from their teachers that many of the kids have older siblings who were with us for this event during one of our three earlier years, and their stories have left high expectations in the minds of today's visitors. Unlike our first year, when we worried over the outcome, now we're confident that this "Meet the Creek" will be as compelling as the others. Our college students will feel humbled and empowered, and the children will feel inspired to learn in this setting where they are free to be wild.

"Meet the Creek" has become a symbolic highlight of our work together during each of the last four years of place-based collaboration. We've been reflecting this year on the beginnings of this undertaking, which has so enriched our professional lives and has had a broad impact on our institution and beyond. In many ways our collaboration was born of frustration as much as inspiration. An English professor by title and a poet and nature writer by avocation, John often feels trapped by the four walls of a typical college classroom. A toxicologist in a former incarnation, Ellen is a biology professor whose teaching load is weighted toward the premedical-minded students in her anatomy and introductory biology classes. Although we have been friends for years, before 2001 we had never worked together in the classroom, as the disciplinary divide between the sciences and the humanities is wide and deep.

Wofford College has a strong academic reputation, and our students are bright eighteen- to twenty-one-year-olds, mostly from the Southeast, but increasingly from around the country. The college's academic calendar reserves the month of January for a compacted term called Interim. This is when the faculty are encouraged to be creative, to accept the risk of failure, even to step out of our disciplines and develop learning opportunities for our students that engage them deeply—whether in some great work of fiction or through some shared experience abroad.

In January 2000, Ellen offered a course titled "Science: What Should Everybody Know?" in which she asked the twenty-one students (mostly freshmen and sophomores) to accept a challenge: "If you partner with me to research and design a new general education course for our non-science majors that addresses the title's question, I'll seek the funding to implement it." For the first time she found herself teaching without a "script" of well-prepared lectures. The students were tentative at first, awkward in this role of partnering with a professor to develop curriculum. It was new territory for all. The Interim team researched the undergraduate science education literature, developed pedagogical goals, brainstormed ideas for course content, benchmarked courses at other institutions, and interviewed administrators, family members, and faculty members for their ideas.

The day John visited the class, however, the students began to sense the possibilities that lay waiting in a different approach. "Why does the question 'What should everyone know about science?' have to be so narrow?" he asked. "It's too confined within disciplinary boundaries. What if, for example, Ellen and I were to blend our freshman courses around some environmental topic broad enough to cross the disciplinary divide? Ellen would bring her knowledge and training as a scientist, a toxicologist, and I would bring my writer's eye and my knowledge of the natural history of the South Carolina piedmont." It was immediately clear to all of us that in such a setting both students and professors could benefit from a deeper understanding of information that different disciplines would bring to a topic. The students were soon buzzing with ideas about other faculty members that they would like to see similarly partnered around a rapidly expanding list of topics.

These ideas for interdisciplinary teaching became the basis for a successful grant application to the National Science Foundation (NSF), titled "Seeing the Big Picture: Linking the Sciences and Humanities."[1] What began as one professor's desire to collaborate with students on developing a new course evolved into an institution-wide initiative to create fully integrated, team-taught, two-course learning communities (LC) serving freshmen in their first semester. The NSF approved its maximum funding ($200,000) to launch the initiative, much of which paid summer stipends for each team of two faculty members (one from sciences and one from humanities) and two students (called preceptors) to work together to develop their LC. The titles of a few LCs suggest the range of topics that emerged, such as "Madness, Creativity, and Literature," "Cosmology and Ultimate Questions," and "Science and Science Fiction." To date eighteen of approximately ninety full-time faculty members have taught in learning communities, which have continued even since NSF funding formally ended in 2003.

While awaiting NSF funding, we launched our pilot learning community in 2001, "The Nature and Culture of Water," for which sixteen freshman students were enrolled. Financial backing for our work came from a unique source—the Spartanburg Water System and Sanitary Sewer District (SWSSSD), whose visionary general manager, Graham Rich, saw our work as directly supporting the utility's public education initiative.[2] Each year, Graham and Ed Neal, the plant manager, give our students a remarkable behind-the-scenes tour of Spartanburg's sewage treatment plant, and the students learn of the plant's codependency with Lawson's Fork, which receives the plant's effluent, and the plant operators' commitment to keeping it clean. "The guys that run this plant are unrecognized heroes," one of our students wrote just weeks after 9/11—but that's another story.

Working with local schoolchildren allows our students to apply what they have been learning in a way that serves the larger community. We prepare for "Meet the Creek" through the first month of the semester. A montage of readings on the library's electronic reserve supplements our fieldwork—there is no textbook for the work we're doing. We begin by introducing the students to our collaborative teaching initiative by invoking C. P. Snow's "Two Cultures." In the discussion that ensues, students recognize that the two cultures at Wofford College are embodied in the "science" (BS) and "non-science" (BA) tracks. We ask them to give us contrasting adjectives for scientists and literary humanists; "Male vs. female" a student says, and we laugh with them about how the two of us reverse that stereotype. We discuss how rhetoric functions in the writings of scientists and humanists, and after their close reading of "Rivers of Death," a chapter of *Silent Spring,* the students begin to see how Rachel Carson's deep understanding of science informs her passion for saving the planet.

By the time "Meet the Creek" comes around in October, our freshmen have studied the stream's ecology and geology, sampled macro invertebrates as indicators of water quality, and learned the science behind the treatment of urban sewage and maintenance of a safe drinking water supply. At the same time they have learned about the stream's natural history, as well as human uses of the stream over the last eight-thousand years. They have written poetry and reflective journals about the stream, which consider such issues as its aesthetic value, its human impact, and whether or not they have any responsibility in protecting and restoring it. In their frustrated struggle to sort the notes in their field notebooks by discipline, students sometimes complain that they don't know when they're learning "science" and when they're learning "humanities." As professors, we're pleased with this indication that we've achieved a true integration of our courses. "Wow," we say to each other with a grin, "it's working!" In another few weeks we'll hear them criticizing the unnatural disciplinary separation that exists in their other courses.

A couple of the elementary school teachers spend several hours with us in September teaching our freshmen about South Carolina's fifth-grade educational standards, and they warn the group that the lessons of "Meet the Creek" must purposefully address those standards. They also stress safety issues and the need for regular "potty breaks." Their underlying message is clear: "We are entrusting our jobs and these children to you for the day, so don't screw up." They also inspire the group by telling us that they often refer back to "Meet the Creek" as they introduce related topics throughout the year, that children continue to talk about the experience weeks later, and that at the end of the year the kids identify their day at Lawson's Fork as the "best thing they did all year."

After the teachers' visit, we share the pictures of scenes from previous years. Then our freshman get down to business by forming four teams, each of which begins the two weeks of brainstorming, planning, and practicing their highly interactive learning station along the stream.

It's now 8:45 A.M., and the freshmen are working to make sure their stations are ready. At one station there will be a scavenger hunt, in which children will find different objects and classify them as biotic or abiotic. Charles, one of the Wofford students, has actually dressed up like a white-tailed deer, with a tanned-hide cape on his back and antlers on his head, to make "biotic" more fun.

At another station freshmen will lead the children to identify different tree species in the riparian zone, and then help each child use paint and rollers to stencil real leaf patterns onto a new white T-shirt that the child takes home. In final preparation, two of our freshmen are sorting the shirts into stacks by size (each child's name tag will have the requested size on the back), while the other two gather leaves from nearby sycamores, mimosas, and water oaks. The tarps are laid out and the plastic bottles of paint are lined up next to the roller trays. They're ready.

At a third station the children will write nature-inspired poetry. The freshmen in this group are testing out their props and practicing their opening "lecture" one more time. Now they're the teachers, building upon the poetry-writing exercise that John had them complete just a couple of weeks ago. They're holding up their posters with funny pictures and glued-on objects that define personification, metaphor, and simile: "The tree's arms reach to the sky"; "A magnolia leaf is like an umbrella." They've set up dozens of colored markers, and the paper they found for the children's compositions is colored like grass. Initially, this will be the least popular station, but it will become a favorite when the poems are read aloud, with much fanfare, by the college students.

The most popular station of all is the macro invertebrate sampling station, where, every forty minutes, fifteen more kids will get to wade into Lawson's Fork and catch, sort, and identify creatures they didn't know they shared the planet with. Several Wofford football players happen to be manning the sampling station this year, and they've squeezed into their waders and are fiddling with their gear—kick nets, buckets, ice trays, tweezers, and plastic spoons. We've watched this day unfold for three years now, and we know that at the end of the day there will be sixty wet and happy fifth graders. We also know how exhausted our own students will be—especially this last group, whose physical activity will have equaled a day of football conditioning.

It would not be completely honest to say that our experiment with outdoor education and disciplinary integration has been universally praised or accepted by all our colleagues. The resistance, often subtle, has sometimes surprised or disheartened us. One humanities professor was opposed to our calling our all-day Thursday outings "field trips," as if such a term would erode the rigor of the Wofford curriculum, suggesting more fun than education; so we adopted the term "field experiences" and continue to unabashedly support combining fun with learning. Another colleague in the sciences worried openly that the freshmen would not learn enough about the scientific method, so we attempted to ease that concern through extensive internal and external assessment of the program.

But in other ways we've been inspired by the widespread support that the learning communities have received. The students are the most outspoken advocates for the program; for example, a formal presentation to the board of trustees by the LCs' preceptors gained a line item of funding to continue the program. We've presented our "Nature and Culture of Water" model at various national conferences and conducted workshops at campuses all around the country.[3] The message seems to resonate widely. People are struck by the common sense of it all: if we respect each other's disciplines, work to integrate our courses, and model for our students a desire to learn from each other, we can't help but transform our students and become better teachers ourselves. And if we empower our students with place-based knowledge and the responsibility to share that knowledge through service-learning projects such as "Meet the Creek," not only are they more cognizant of the natural world that surrounds them, but they are also more responsible toward the human community that exists outside the walls of Wofford.

A little after nine o'clock the yellow school bus finally pulls into the drive, and Ellen yells to all, "They're here!" Bea Bruce, the Chapman Elementary librarian, steps out of her car, which precedes the bus. Bea has been our contact at Chapman all these years; today she laughs, admitting that someone forgot to reserve the buses in advance, but she scrambled to find a driver and squeezed all the kids onto one bus, and somehow they still managed to show up on time. Our freshmen swallow their trepidation and greet the excited kids as they spill out of the bus. Each child is wearing a name tag in one of four shapes—blue dolphin, green frog, red leaf, or yellow butterfly—so that we can keep them sorted into four equal groups.

Ellen, the pied piper of the event, welcomes everyone, toots her famous air horn, and guides the group to the lawn behind the clubhouse to begin the day. She leads them in a call and response to get their blood moving: "Which is the BEST elementary school in Spartanburg?" "Chapman!" they scream louder and louder as Ellen repeats the question several times. "What's the name of our college?" "Wofford!" they yell, and we break up into the four groups.

Throughout the morning, each jostling, silly group of kids moves from one station to the next—the dolphins go first to the leaf / T-shirt station, the frogs to the scavenger hunt, the red leaves to poetry, and the butterflies to get wet in the creek. It is a noisy swirl of joy and learning that rotates through the different stops every forty minutes when the air horn blows. The teachers watch, helping out if needed, but glad to have a well-deserved day off from instructing. The Wofford students take pride in their teaching, and the elementary teachers are surprised to see many of their most problematic charges become stars in this setting. Last year after "Meet the Creek," one boy "who had never written more than two sentences" filled up two pages with his excited description of the day. His teacher was so amazed that she sent it to us to share with the class.

The kids learn that leaves have veins like we do, and they stencil those same leaves onto their new white T-shirts. They scramble and dig around in the sand, chanting "biotic" or "abiotic" as they pick up scraps of wood or debris left behind by last week's flood. The young poets wave their poems in the air to ensure that they are read aloud by their new heroes from Wofford College. The football players in waders lead another group into the creek's cold October water, and the kids squeal as water and sand fill their shoes. The young men hold the kick nets on the stream bottom as the kids dance in front of them to dislodge the critters and send them into the nets. Then the nets are held up and little heads excitedly crowd around and peer in. With each exclamation of "There's one!" a child eagerly dips in with a plastic spoon for a prize: another dragonfly or mayfly larva. "Lobster!" they yell with each crayfish that is caught, and they're particularly thrilled with one small catfish (not an invertebrate, we have to point out to them over and over). If they'd netted a shark they wouldn't have been any more excited.

The cool morning gives way to a warmer midday, and we stop for lunch. The kids sit with the Wofford students, and laugh and tell stories from the three stations they've been through this morning. The one dry group is now glad that they will get to go into the stream last. After one more round, the happy, tired, dirty, and wet kids collect their T-shirts from the hangers suspended on a temporary clothesline. We expect them to head straight to the bus, but a request from one child for the football players' autographs turns into a chaotic assembly line where our young men patiently sign their names on shirts. Finally satisfied, the children pile back onto the bus and head away, ultimately going back home to where some have their own creeks. "I'm going home today and look at the creek behind my house," one little boy says. "I'll bet there's something living in there, too."

Once again we've done what we came to do—give sixty local kids the chance to, as environmental artist Andy Goldsworthy describes it, "shake hands" with this place we call Lawson's Fork. From the look of their clothes, some have hugged it and sat in its lap! Now we can pack up and go, too, but first we all make our own leaf-stenciled T-shirts, and it gives us a chance to share the day's highlights. Our college women laugh about the number of young boys who asked them out on dates, and several of us recount favorite phrases from the children's poetry. "The mosquitoes are like my sister and brother" generates a chuckle. One entire poem is embedded in our memories:

> The waterfall is like a child crying over a piece of candy.
> The smell is like a clear summer day when the wind blows.
> When the water hits the rocks it goes splash, splash, kapow.
> It sounds like milk splashing down in the cheerio bowl.

We're all filled with pride at what we've accomplished today, and the bonds developed from this shared experience will hold us close for the rest of the semester and beyond. "Meet the Creek" symbolizes the reason why the two of us added this work to already heavy teaching loads, and once again a deep feeling of fulfillment washes over us.

NOTES

1. For more information about the NSF grant and the LC program at Wofford College, please contact Ellen Goldey (goldeyes@wofford.edu) or go to http://www.aaas.org/publications/books_reports/CCLI/pdfs/07_imc_Goldey.pdf to download the following book chapter: Ellen Goldey, "Disciplinary Integration: The Sciences and Humanities in Learning Communities," in *Invention and Impact: Building Excellence in Undergraduate Science, Technology, Engineering, and Mathematics (STEM) Education*, 209–15 (Washington, D.C.: AAAS Publications. 2004).

2. SWSSSD has continued to support our work and the dissemination of its message with grants totaling $45,000, and we deeply appreciate this college/community partnership.

3. We have given workshops and presentations on our LC model at the SENCER (Science Education for New Civic Engagements and Responsibilities) Summer Institutes (www.sencer.net) and at conferences sponsored by the American Association of Colleges and Universities, the National Learning Communities Project, and the Association for the Study of Literature and Environment. We even presented with SWSSSD's Graham Rich at the 75th Annual Water Environment Federation Conference and Exposition. We've also led workshops for faculty at several institutions, including Franklin-Pierce College (N.H.), St. Mary's College (Calif.), and the University of Wisconsin–La Crosse.

COURSE READINGS

This list represents a sample of required readings (available to students through college library e-reserve) used in the five weeks leading up to "Meet the Creek."

Kaufman, Donald G., and Cecilia M. Franz. "Science and Environmental Studies," chapter 2, and "Ecosystem Structure," chapter 3, in *Biosphere 2000: Protecting Our Global Environment*. 3rd ed. Dubuque, Iowa: Kendall/Hunt Publishing, 1999.

Keller, Edward A. "Rivers and Flooding," chapter 6, in *Introduction to Environmental Geology*. 3rd ed. Upper Saddle River, N.J.: Prentice Hall.

Lopez, Barry. "Children in the Woods," in *Crossing Open Ground* (New York: Vintage, 1989).

O'Conner, Flannery. "The River," in *The Complete Stories* (New York: Farrar, Straus, and Giroux, 1996).

Orr, David. "Reflections on Water and Oil," in *Earth in Mind: On Education, Environment, and Human Prospect* (Washington, D.C.: Island Press, 1994).

Ray, Janisse, and David Scott. Watershed Journal, http://www.lawsonsfork.org/journal /Journal.html.

Snow, C. P. *The Two Cultures* (Cambridge: Cambridge University Press, 1998).

South Carolina's Fifth Grade Educational Standards. http://www.sceoc.org/guides-to-scc-standards.htm.

Taylor, David, and Gary Henderson. *The Lawson's Fork: Headwaters to Confluence.* Hub City Writers Project, 2000.

CHAPTER 9

Beneath the Surface

Natural Landscapes, Cultural Meanings, and Teaching About Place

KENT C. RYDEN

In one way or another, almost all of my pedagogical activity involves teaching about place. The more I think about it, though, the more I realize that what's really going on is that I'm allowing places to teach me. And to be a good student of place, I've learned, you have to be a careful reader and listener. It's easy to be beguiled by the immediate qualities of an attractive landscape. But we need to look beyond those qualities to the many other stories that may be hidden there in order to understand just how rich a resource places can be for teaching and learning, and just how important their lessons can be.

Let me explain. I've always been drawn to places that have complicated stories to tell. Indeed, I think the reason that I've followed this professional path is that I've always lived in places that have given me plenty to think about. I spent half my childhood in western Connecticut, and followed that up with chunks of adulthood in central Rhode Island and southern Maine. All three of these places offered me plenty of woods to tramp around in—one of my favorite ways to spend an afternoon or a day—and cemented my love of being in the natural world. And yet these were

New England woods, and New England woods always try to teach you more than meets the eye if you know how to look at them the right way. I never had to walk far in my nearby woods before I came across a stone wall or an old foundation; my favorite swimming hole in Connecticut backed up behind the dam of a long-gone paper mill. My natural haunts were full of human artifacts, evidence that many people had been there long before me trying to make the landscape produce useful things for them, be it crops or hay or water power. The places that I enjoyed for their green and watery qualities, I came to understand, looked that way only because of things that farmers and builders had done there in the past.

As I've continued to read and travel and walk around over the years, I've realized that the lessons I've learned in New England can be applied anywhere. Any landscape—even one we think of as epitomizing "nature"—is a kind of historical artifact, the end result of natural and cultural processes working together. This hasn't affected my love of the New England woods. Far from it. As with the books I read, I like my landscapes to be challenging and thought-provoking. But I've also picked up a lesson that's important for anyone who teaches about place. Even as we instill a love and understanding of the natural world, we don't tell the whole story if we go no deeper than the leafy and wooded surface. And the more of the story we tell, the better and more responsible our teaching can be: if we understand that places look the way they do because of things that people have done in the past, we can think more effectively about what we might do to guide those places into a healthy future. Once we understand that "nature" is not separate from "culture," we have to think more carefully about the roles we must inevitably take in shaping and sustaining the places that we love so well.

I teach in a graduate program in American and New England studies, and thus I've been able to bring many different disciplinary perspectives together when my students and I talk about place. Depending on the class and the particular assignment, we might bring the lenses of literary studies, environmental history, cultural geography, and other related fields to bear as we try to understand what a particular place is trying to teach us. In some places, the many layers of meaning that a place contains stand out with particular clarity, and I find that these places make especially good teaching tools, both for understanding a specific locality and for raising questions about the complex nature of place more broadly. In the summer of 2005, for example, I included Rachel Carson's *Silent Spring* as part of my "Landmarks of American Nature Writing" seminar. On the last day of class, we gathered in the coastal Maine town of Wells to visit the Rachel Carson National Wildlife Refuge, a place with many preexisting meanings that visitors have to encounter before they walk through the refuge itself. Their vision is influenced by the historical reputation of Rachel Carson, in some cases by her writings (particularly *Silent Spring*), by the official endorsement of the U.S. Fish and Wildlife Service, and by the very notion of a "wildlife refuge" and the implications that such a term suggests. Once visitors enter the refuge, they are admonished to stay on a wide and well-maintained path, provided with scenic outlooks that focus their attention in certain directions for certain teaching purposes, and encouraged to bring along an interpretive trail guide and an explanatory pamphlet that nudge them toward grasping what the Fish and Wildlife Service explicitly wants them to think the place means.

And it means many things at once. The refuge is at the same time a coastal Maine salt marsh, a piece of government property, a symbol of

American attitudes toward the natural world, a physical reminder of a particular writer's books, and a stop on a Maine tourist itinerary that helps people understand, as our state slogan puts it, "the way life should be." Some of these identities overlap, some contradict each other. As my students and I walked through the refuge together, thinking about the literature class we had just finished, reading the lessons that the landscape had been specifically shaped to tell us, talking about the history of "nature" in American culture, we appreciated the thick bundle of meanings that the place made available for us to tease apart, meanings that were both grounded in a specific spot in coastal Maine and pointed to much larger issues of how to interpret places in general. The Rachel Carson National Wildlife Refuge makes the process fairly easy, but it also suggests questions that we can ask wherever we go, in landscapes that aren't helpfully labeled for us and where the trail guides are drawn up by our own whims.

While Rachel Carson's books about the ocean and the life it contains— *The Sea Around Us* and *The Edge of the Sea*—were quite popular when first published in the 1950s, she is probably best known today as the author of *Silent Spring,* a book that many historians credit with helping to curb the widespread use of pesticides, particularly DDT, in American agriculture. Carson was a sometime resident of Southport Island in Maine, and so it seems entirely appropriate to honor her with a wildlife refuge in the state that she loved. At the same time, *Silent Spring* is not so much a book about preserving nature as it is about safeguarding the health of both human and animal populations in the places that they share. Carson's view, like the one I developed in the New England woods, is of a world in which the natural and the cultural are tightly entwined; she begins her book with a nightmare "Fable for Tomorrow," in which "a town in the heart of America where all life seemed to live in harmony with its surroundings" suffers seemingly inexplicable sickness and death among human, bird, and plant populations. While Carson supported the preservation of nature, her most famous and influential book locates nature in towns, in backyards, in our own bodies. Nature lives where we do, rather than out there somewhere beyond where the pavement ends.

The class and I had noted and discussed Carson's view of nature, and we had also talked about its difference from the versions of wild nature that emerge from the pages of writers from John Muir to Edward Abbey. These were the literary lenses that we brought with us to the refuge, and thus we were immediately struck by how the landscape of the place presented us with an understanding of nature that was the exact opposite of the one that Carson worked out in the pages of her book. The concept of nature that we saw in the refuge was one that draws and polices a firm line between the world of nature and the world of humans rather than a single world of mutually involved parts. Even the idea of a "wildlife refuge" is telling, as it evokes the image of a secure hiding place where wild animals can be safe from the nasty humans outside the gates; it is a landscape that is "theirs" as opposed to "ours." Humans may enter, but only on certain restrictive terms. The class and I were struck by the ways in which our experience in the refuge resembled a visit to a museum, where attractive specimens of high-quality nature are displayed for our viewing pleasure but always kept behind figurative velvet ropes.

An information kiosk at the entrance to the refuge's walking trail (supplemented by the official Rachel Carson National Wildlife Refuge brochure and a separate pamphlet on the refuge's seasonal bird populations) touted the wonders within: white-tailed deer, river otter, beaver, fox, coyote, moose, harbor seals, turtles and frogs, and scores of bird species ranked by their "abundant," "common," "uncommon," "occasional," or "rare" appearances. To our amusement, we didn't see anything as interesting and exciting as a moose or a seal; I think we topped out at a snowy egret or two. Then again, we weren't able to search extensively for the foxes and otters that might have been hiding from us, for we were admonished by signs and brochures to stay on the trail to avoid damaging vegetation and disturbing wildlife. A sensible precaution, to be sure, but one that heightened our sense that we were in a place where we were alien intruders and where nature had to be admired from a distance, since we would only mess up if we actually came in contact with it. If wildlife deigned to show itself to us, we could count ourselves lucky.

The refuge provides a trail guide to the one-mile loop that winds through its woods and along the edge of its salt marsh, informing visitors about the interactions of land and water that characterize this particular location, and about the flora and fauna that occupy its various microhabitats. Anyone reading the guide thus gains a sense of the ecological complexity of the refuge's landscape, and yet that complexity is simplified by the museum-like quality of the trail, which reduces an interdependent mix of land, water, and wildlife to a series of pretty scenes. The guide brings viewers to a sequence of numbered vantage points, sometimes enhanced by wooden observation decks, where they have presented to them a view titled "The Edge of the Marsh," say, or "Hemlock Hollow," "The Tidal Flux," "Meanderings"—titles that suggest paintings, and therefore perpetuate the idea that nature here is fundamentally a static and visual thing, not a bundle of vigorous natural processes or something in which humans participate. Indeed, some of the explanations in the trail guide serve as figurative "Do Not Touch" signs, as when we learn at "The Edge of the Marsh" that "undisturbed coastal wetlands that have natural vegetation along their edges produce dense meadows of grasses and other plants that support abundant wild-life. When marsh edges are cleared for buildings or otherwise disturbed, use of the marsh by wildlife declines." Moreover, by highlighting eleven discrete spots in the refuge, the trail and its guide suggest that the segments of landscape between the numbered views are unimportant, not as interesting and meaningful as the places where we are told to stop and look and learn, an impression that again obscures any sense of the refuge landscape as a dynamic and seamless whole. Instead, it is a series of pictures in an exhibition; no one remembers what the floors look like in the Louvre.

The class appreciated these ironies, joking that we shouldn't crane our necks too far while enjoying the views lest we break out of the frame that the trail's designers intended us to see through. This pointed to a further irony: the fact that our experience of "undisturbed" coastal scenes was accomplished through a lot of human work, as we were led through the refuge on a carefully designed and maintained trail, had our eyes pointed in certain directions, and were instructed in what to think about what we saw. It's the sort of landscape that tries to deny its own human-made quality, though, in that you are encouraged to enjoy scenes of natural beauty without thinking about how you got there. The cultural history of the landscape is hidden in plain sight, as it were. The Rachel Carson National Wildlife Refuge is itself a humanly created thing, of course. It is a bounded property on the map that separates a "wild" world from a tamed one; it is a category of federally protected and managed property; and the simple measure of deciding that something is "wild" is itself a cultural act. The refuge's official brochure does have a page acknowledging that the southern Maine coast has a human history, which seems to break down into four periods: Native American settlements, colonial occupation, shipbuilding and fishing in the 1800s, and tourism and recreational use from the mid-nineteenth century ("Interest and access were particularly spurred by the arrival of the railroad in 1842") to the present. Extractive uses of the landscape stopped long ago, it seems, and the coast has been able to be "nature" ever since, embraced by urban tourists when they need a refreshing break. At any rate, none of this sketchy history—Native American sites, say, or evidence of building and launching ships—is visible from the refuge's trail. The human world stops, it seems, when we turn off of Route 9 and walk away from the parking lot into the woods.

One of the dangers of being a teacher is that sometimes people think you make things too complicated. This is the point at which a student usually cocks an eyebrow at me and says, "Oh, come on. Aren't you being a little too cynical? Don't wildlife refuges do a lot of good regardless of what they make us think about nature?" Of course they do. I am all for protecting habitat wherever it may be, and a place like the Carson refuge can teach people a lot about how salt marshes work and the role they play in broad cycles of plant and animal life. The refuge is becoming more and more an isolated island of "nature" along the rapidly developing southern Maine coast, but the encroaching tides of "culture" make its presence that much more vital. Still, one of the things that my university pays me for is to be a contrarian (or so I like to believe), and I feel obliged to teach my students to think critically about places rather than taking them at face value or ignoring them completely.

The Carson refuge amply repaid our visit because it clearly reveals many facets of meaning at once. We approached it through the literary overlay of Rachel Carson's writings and the environmentalist connotations of the name "Rachel Carson" itself. Once there, we encountered a carefully shaped landscape that conveyed a somewhat different message than that delivered by *Silent Spring*. Rather than encourage us to be stewards of the nature that surrounds our daily lives, the actual message that was communicated reinforced some common American cultural assumptions about nature as a place apart, one where humans can visit but never live. That is, the physical landscape of the refuge also connects to much broader trends in American cultural history, trends that run through the works of a John Muir and an Edward Abbey on the one hand to the national parks and the popularity of ecotourism on the other. It is difficult not to bring this larger framework to bear when encountering the refuge landscape and considering its meanings, and so the Carson refuge in the end is both a specific concrete spot of earth and a summary of an entire world of literary and cultural history. When we as a class set out to read this landscape as a communicative text, we found that the trail guide was only the most immediate of many guides that we had available to interpret what we saw.

As I have said, the Carson refuge made this process easy. It is an obviously authored place, one that tries to give you a specific message both through written text, in the form of signage and brochures, and through the carefully crafted shape of the landscape itself. It is difficult to finish a walk on the trail and not come away with a sense that you have just seen an official version of "Nature"—a nature, moreover, that is comfortably recognizable, filled with special things of the sort that you can see only when you visit a place like this. The messages that the Carson refuge intentionally beams to you are not particularly challenging or complicated. Still, once you become conscious of the way that the refuge works to shape your thought, you realize that you can look beyond those meanings through methods and habits of seeing and questioning that you can bring to any landscape, any place, even those that are not so obviously authored. Not all landscapes guide you through on a trail that you are forbidden to leave, but they nevertheless shape and constrain our movements, putting us in certain relationships with nature and other people, and in that way send subtle messages about the expected relationship between the two. Moreover, as I learned in my New England woods, every place is an environmental text, from a second-growth forest to a wildlife refuge to the residential landscapes where we all live. All acts of building (or choosing not to build), from houses to roads to fences to farms to cities, grow out of an environmental ethic, a set of assumptions about what is right, proper, and necessary to do to the environments in which we find ourselves. When we read places carefully, then, we see how they make visible the social, cultural, economic, and environmental histories and priorities of the people who make them and dwell in them.

It's not always easy to tease out these many levels of significance, particularly when you're walking down the street, as opposed to visiting a place like the Carson refuge. This is where being interdisciplinary comes in handy. As a folklorist, I have gathered stories from residents of a north Idaho mining district; while the mines tell a story of human dominance over nature and its resources, patterns of local narrative reveal a place whose residents see it in much more ambivalent terms. As a cultural geographer, I know how to see landscapes as texts, as material culture that grows out of human minds as well as human wills, as sets of meaningful objects as well as practical artifacts that keep the rain off our heads and let us move from one place to another. As a historian, I put the environmental assumptions that words and things embody into a larger context, linking the thoughts about nature that are symbolized around me with the thoughts of people in other times and places. And as a teacher, I try to get my students to see and understand in these ways as well, to find in even the most everyday scenes an eloquent testimony to the thoughts and acts of the people who have made them, who have used them, and who live and move in them today. Every piece of evidence tells another part of the story. And as I said before, I think it's important that we refuse to let our students complacently take places at face value; rather, we must encourage them to try to understand the many cultural meanings and historical tales that those places are waiting to teach them. Of course, I don't expect that other teachers will be able to take in large bodies of scholarship from disciplines outside their own in order to be able to teach about place in this broadly interdisciplinary way. Still, if we can at least suggest to students the richness of the places they see every day, the officially sanctioned and clearly authored ones as well as the obscure and mundane ones, we can increase their pleasure in their experience of the world and, if we're lucky, make them more responsible human beings as well.

How? Quite simply, every place is meaningful. The world is not a blank globe on which we can do whatever we want; as I began learning at an early age, every place bears stories, memories, the physical leavings of people long gone. I would hope that no one would want to bulldoze the Rachel Carson National Wildlife Refuge and turn it over to the highest bidder. When we think about it, though, every place is its own version of the Carson refuge, which may not stand for "Nature" so clearly, may not seem to stand for nature in any obvious way at all—think of your average urban city block—and yet it bears a layer of imaginative, social, cultural, and historical significance all its own. There is much to disturb at the refuge, and it's not always animal habitat and fragile ecosystem. Both natural and human ecologies deserve our attention and respect, and we should engage in a process of long and hard thought before we choose to disturb them—if indeed we disturb them at all. And if some places seem, frankly, socially and environmentally hopeless, we can still understand them as bad examples so that we can collectively do better in the future. Places have much to teach us; indeed, they teach us every day without our necessarily realizing it. And the more we know—the more questions we can ask from more perspectives—the more we learn.

COURSE READINGS

Asterisked entries are from the *Norton Book of Nature Writing,* edited by Robert Finch and John Elder. New York: Norton, 2002.

Abbey, Edward. *Desert Solitaire.*
*Audubon, John James. *Ornithological Biography,* excerpt.
Austin, Mary. *The Land of Little Rain.*
Bartram, William. *Travels.*
Carson, Rachel. *Silent Spring.*
*Catlin, George. *Letters and Notes on the Manners, Customs, and Conditions of the North American Indians,* excerpt.
Dillard, Annie. *Pilgrim at Tinker Creek.*
*Eiseley, Loren. "The Judgment of the Birds" and "The Star Thrower."
Emerson, Ralph Waldo. "Nature."
*Hoagland, Edward. "Hailing the Elusory Mountain Lion."
*King, Clarence. *Mountaineering in the Sierra Nevada,* excerpt.
*Lopez, Barry. "The Raven."
Muir, John. *The Mountains of California.*
*Powell, John Wesley. *Exploration of the Colorado River,* excerpt.
Snyder, Gary. *The Practice of the Wild.*
Thoreau, Henry David. "Walking."

PART III ✍ MEETING THE CHALLENGES

CHAPTER 10

Idiot Out Wandering Around
A Few Words About Teaching Place
in the Heartland

JOHN PRICE

The first time I encountered this phrase, about ten years ago, I was visiting the writer Dan O'Brien at his ranch in western South Dakota. Dan had generously offered to drive me around the surrounding acres of mixed-grass prairie that he was restoring and to hunt ducks with his peregrine falcon, Little Bird—one of a native species he had helped save from extinction. Though I'd been born and raised in central Iowa, I knew virtually nothing about prairie ecosystems. I grew up as a townie, surrounded by an ocean of corn and soybeans, in a state where less than one-tenth of one percent of native habitat remains—the worst percentage in the Union. Native grasses and flowers might have been found in the place where I grew up, in the occasional ditch and postage-stamp preserve, but I would not have been able to identify them. Even if I could have, I'm not sure I would've cared.

"Not caring" was just one of the conditions I was hoping to alleviate by traveling to significant prairie sites in the region and talking with writers, like Dan, who were dedicated to preserving and restoring them. Altogether, those journeys would transform my relationship to place, helping, as I later wrote, "to cure a lifetime of ignorance and indifference." This personal transformation—or more accurately, conversion—would be the subject of my first book and would lead me to become a committed writer of place in the central grasslands. At the time I visited Dan, however, that commitment was tentative, and my lack of any significant outdoor experience didn't help. During my most recent visit to Dan's ranch, for example, I'd nearly fallen off a horse while galloping across the open and (for me) surprisingly un-flat prairie. But that was all in the past, I thought, as Dan slowly parked the truck near a pond, pointing to a handful of ducks floating on the water. It was the perfect situation to see Little Bird hunt, to experience at least a faint echo of the once grander cycles of predator and prey in the region. So I lost myself for a moment—as I often did on that grasslands pilgrimage—and after stepping out of the truck, slammed the door behind me, the noise reverberating across the pond like a shotgun blast. Luckily, it didn't scare the ducks off, but I was still embarrassed when I met up with Dan at the back of the truck.

"You know what 'Iowa' stands for, don't you?" he asked me, smiling. "'Idiots Out Wandering Around.'"

During the intervening years, I have done my best to escape this label, but to no avail. I now teach at the metropolitan University of Nebraska at Omaha, and a few years ago, during the first meeting of an introductory lit class, I casually mentioned that I grew up in Iowa and still lived there. This provoked some snickering. Given that Omaha and Iowa are separated only by the Missouri River, I was curious to know what they saw as the main distinction. "What is an Iowan to you?" I asked. This provoked the Idiot line and another local favorite: In Omaha Without Authorization. Then I asked them what they thought our two states had in common. Most of the positive responses I received can be summarized by the popular term "Heartland Pride": family values, simple country living, self-reliance, hard workers, agriculture, patriotism, and God. The negative responses can be summarized by what might be called "Heartland Funk": ignorant, ugly, culturally and environmentally boring, and best left behind. There wasn't a single reference to a natural feature unrelated to agriculture, not even the Mighty Mo. Their responses were yet another indication of the severity of the situation in the Great Plains and prairie Midwest, where the vast majority of native habitats have been destroyed or are underprotected. The consequence, as writer and ecologist Dan Flores has said, is that "citizens of places like Texas and Kansas are today among the most divorced of all Americans from any kind of connection with regional nature" (15).

I should say up front that I have encountered many students here who know a lot more than I do about local ecosystems and who are dedicated environmental activists. The majority, however, are where I was ten years ago, struggling to become more educated about the bioregion and, perhaps more important, to find a reason to care. To be honest, I'm still struggling, and not the least of that has to do with my role as a teacher. Place is a significant component of all my courses, whether a graduate seminar titled "Grasslands Autobiography" or dual-level courses like "Environmental Literature" and "Great Plains Literature" or nonfiction creative writing courses or introductory lit surveys. I confess, however, that my approach has not been a terribly organized one; it continues to change and adapt as I seek to connect with each individual class and student and season, in my place and in my life. Over the years, however, some guiding principles have emerged, and one of them is that the same regional abstractions and stereotypes that disconnect us from place can also, with a little revision, draw us back in. That's been true for "Idiot Out Wandering Around," which has provided me with a helpful mnemonic for organizing a few of my ideas, experiences, struggles, and hopes as a teacher of place.

IDIOT

The destruction of native grasslands provides a powerful example of the ways in which uninformed writing adversely affects place. The aesthetic denigration of the grasslands is nothing new—it can be found in works by some of the world's most popular authors. Charles Dickens, while visiting Illinois' Looking-Glass Prairie in 1842, declared it "the great blank," not to be remembered "with much pleasure, or to covet the looking-on again in after life" (182–83). James Fenimore Cooper, who never visited the trans-Missouri grasslands, nevertheless saw fit to have his famous character, Natty Bumpo (or Hawkeye), proclaim: "I often think the Lord has placed this barren belt of prairie behind the states to warn men to what their folly may yet bring the land!" (24). The process of literary canonization helped entrench these negative opinions, validating efforts to convert prairie into forest or field, while neglecting opposing views expressed in early travelogues and Native American oral tales. Perspectives such as Albert Pike's, who in his *Journeys in the Prairie* (1831–32) wrote: "The sea, the woods, the mountains, all suffer in comparison with the prairie. . . . Its sublimity arises from its unbounded extent, its barren monotony and desolation, its still, unmoved, calm stern, almost self-confident grandeur, its strange power of deception, its want of echo, and, in fine, its power of throwing a man back upon himself" (quoted in Flores, 3). Or the Irish explorer Sam Butler, who in the 1870s wrote that the "great ocean itself does not present more infinite variety than does this prairie-ocean of which we speak" (quoted in Gayton, 97). Power? Grandeur? Variety? In nineteenth-century Iowa, when leaders had an opportunity to embrace this remarkable natural heritage, they decided instead to nickname their new state after Hawkeye, the popular literary character who loathed it.

"Idiots indeed," a student once remarked when I mentioned this in class. But as we discuss others' ignorance of place, we inevitably come back around to our own. Together, we talk about what we know about local ecosystems and what we don't know; what we think is here, what is actually here, and what we wish were here; what we love and what we hate about living in this place. These discussions can become digressive, heated, embarrassing, humorous, and ultimately unsatisfying, but they can also set the stage for more formal explorations of the subject in personal essays—a form I assign in almost every class I teach. The personal essay, along with much of "creative nonfiction," is frequently dismissed as easy to write or as a solipsistic exercise in "licking the mirror," as Bill Maher once put it on his television show. A successful personal essay, however, is never simply about the self; its goal is ultimately to discover and articulate relationship. Writing personal essays can challenge students, through reflection and research, to confront contradiction, conflict, and complexity in their relationship to place and then explore the consequences. Doing so is essential in a bioregion where the oversimplification and destruction of native habitats has been intimately entwined with the oversimplification of the self.

Writing and studying the personal essay can also encourage students to cross the often intimidating border between the humanities and the sciences, and to practice the kinds of interdisciplinary research that most agree is essential to any well-rounded understanding of place. Nature writer Chris Anderson argues that a spirit of flawed discovery guides the essay form, which comes from the French for "to attempt" or "to try," and also scientific research, which, according to biologist and essayist Lewis Thomas, "is dependent on the human capacity for making predictions that are wrong, and on the even more human gift for bouncing back to try again" (quoted in Anderson, 316). Both scientists and essayists, according to Anderson, value "error, ambiguity, and the peripheral" as essential modes of discovery. This is certainly true in the work of my favorite contemporary grasslands essayists (all of whom I've assigned in various courses), including Linda Hasselstrom, Dan O'Brien, William Least Heat-Moon, Mary Swander, Scott Russell Sanders, Julene Bair, and the late Paul Gruchow. Perhaps this is because the prairie places they've written about have been so severely affected by scientific error, ecological marginalization, and public ambiguity over their protection and restoration.

OUT

Like many of my students, I do not have an instinctive love for spending large amounts of time outdoors. The sources of this are at once personal and bioregional. My family did not hunt or fish or own a canoe or a summer lake home, and though my mother forced me to join the Boy Scouts I was never, if you'll excuse the expression, a happy camper. At first my wife, Stephanie, who spent many blissful summers camping in the Sawtooth wilderness areas near her Idaho home, found this mystifying. Now that she lives here, however, she appreciates the local challenges: the crowds, the humidity, the hailstorms, the swarms of mosquitoes that may or may not infect you with West Nile virus—and the lack of any large wilderness areas to make the effort seem, on the surface, worthwhile.

This is not to make excuses, I tell my students, but only to articulate yet another kind of place-based ignorance that needs addressing. One way to do this is to take a long journey, as I did, to immerse yourself in a few of the largest remaining prairie sites in the region, perhaps in Kansas or South Dakota. Camp, walk the miles, become vulnerable to your native place. Many nod when I tell them this, but just as many roll their eyes or stare at me blankly. Ours is mostly a commuter campus, which means a lot of my students are holding down full-time jobs, raising children (sometimes alone), struggling with poverty, and/or residing an hour or more from campus. To suggest that they take an extended journey to better appreciate grass is not only impractical for some, or even absurd, it is also insensitive—yet another consequence of privilege. In my own life, as well, personal and professional responsibilities have complicated any easy notions I've had about repeating my sojourn across the grasslands.

"So, okay, what can we do?" I've asked them and, together, in a variety of courses, we've practiced alternative, bioregionally inspired ways to travel and write about place. Similar to a native bluestem, fixed in place, we extend ourselves like roots into the immediate surroundings, seeking connection. And if that doesn't work, we set loose the seeds, our children (or the children of friends), who will quickly find something worthy of attention and, more than likely, concern. We also seek to become mindful of the ways local nature comes to us, compromising established boundaries and taking root in the margins of daily life. Some of the best environmental essays I've received have focused on fraternity refrigerators, wall mice, and office windowsills. Even toilets. In one such essay, submitted during a nonfiction creative writing course, a woman recalls returning from a high school skiing trip to discover a dead rat in her toilet. As it turned out, the demolition of a nearby meatpacking plant had driven rats into a number of south Omaha homes. Writing about this made clear the connection between local economy and ecology, as well as between her life and the seemingly distant lives of her Polish immigrant ancestors who worked in those slaughterhouses. Here, once again, is the bluestem, extending itself vertically, beneath the surface. Here, also, is the essayist of place, "penetrating and describing," as Kent Ryden has said, "these visible and invisible layers of history, communicating the rich physical and imaginative textures of a landscape" (222).

While enriching our relationship to home, the process of researching and writing local essays of place can also be painful, forcing us to revise some of our most cherished regional myths. "Simple country living" becomes something else when a student determines that the quaint abandoned filling station on the edge of a farm town is leaking petroleum into the drinking water. "Family values" become something else when a pregnant student nearly dies from an allergic reaction to the polluting fumes of a nearby steel foundry. "Self-reliance" becomes something else when a student is informed that the soil in her backyard vegetable garden is part of a massive, lead-contaminated Superfund site in eastern Omaha. Long aware of how little some people value their cultures and landscapes, these students and others are still surprised to discover how little some people value their actual lives.

WANDERING

I'm afraid, however, that what students discover about the endangered local environment can become just another excuse to abandon it—"ugly, boring, *and poisonous!*" Those discoveries can also lead us to believe that the most vital battles to save the earth are taking place elsewhere, in bioregions where there is still some wild earth to save. The Loess Hills, however, which run almost the entire length of the Iowa side of the Missouri River, offer a unique opportunity to challenge this myth. The majority of what remains of native Iowa prairie is found in those dramatic, nearly three-hundred-foot-high hills. Created over millennia by windblown, glacial silt, the Loess Hills contain unique communities of native species—including cacti and lizards—and their size and range are matched by only one other loess formation in the world, in China. They are a regional and national treasure, and many of my students know virtually nothing about them. They aren't alone—I grew up just two hours away, and I also knew nothing about them.

So when I can, I take small groups of students to visit the Loess Hills, where we always encounter enough prairie to inspire a flicker of awe, even reverence. While there, we've learned about the rituals of prairie restoration, its absolute dependence on volunteer labor for seed harvesting and planting—a return to some of the low-tech, cooperative techniques used by earlier farmers. We've learned to appreciate the ecological and aesthetic threat of encroaching housing developments. We've also learned to appreciate the ecological and aesthetic threat of encroaching species of trees, and the necessity of clearing them—no small thing given that Arbor Day was founded in Nebraska. We've observed the worst in human treatment—entire hills cut in half for landfill—and also extraordinarily creative responses, such as the "grass bank" program that allows landowners to temporarily move their livestock to healthy prairie while their own land is being restored. The end result of all these efforts remains unclear, as it does for many ongoing prairie restorations. As Pauline Drobney, project botanist at the Neal Smith National Wildlife Refuge in central Iowa, once told me, "[Prairie restoration] is a big riddle, an experiment in the unknown."

Because of this it is also, for many, an exercise of faith—another kind of wandering. Most who participate in restoration will not live to see its ultimate fruition, a fully healed prairie ecosystem. That will take many decades, perhaps centuries. They nevertheless keep working, sustained perhaps by the same inner reserve that nurtures spiritual belief. As historian Donald Worster has pointed out, John Muir, while distancing himself from his Campbellite upbringing in Wisconsin, may have fashioned some of his environmental beliefs and practices after that religious discipline. Likewise, native Iowan and ecologist Aldo Leopold wrote, "We can be ethical only in relation to something we can see, feel, understand, love or otherwise have faith in" (251). Louise Erdrich states, "I would be converted to a religion of grass. . . . *Bow beneath the arm of fire. Connect underground. Provide. Provide. Be lovely and do no harm*" (149). Even parts of the Bible, such as where Paul defines faith as "the substance of things hoped for, the evidence of things not seen," can be reinterpreted from within a bioregional, ecological imagination. Like everything else, spirituality—yet another oversimplified characteristic of the heartland—is not separate from place. Nor should it necessarily be excluded from the process of writing and thinking about environmental literature.

AROUND

I do not hide the fact that one of my greatest hopes as a teacher is that, wherever else my students may roam, they will eventually return home to stay. Grasslands ecologist Wes Jackson argues that our "universities now offer only one serious major: upward mobility. Little attention is paid to educating the young to return home, or to go some other place, and dig in. There is no such thing as a 'homecoming major'" (3). Part of the challenge for my creative writing and literature students is that there are too few examples of "homecoming" in the regional literary canon. Most of the "biggies"—including Mark Twain, Willa Cather, and Hamlin Garland—wrote about their prairie childhoods from their homes in coastal cities. The reasons for this are practical, such as proximity to research libraries and publishers, but they may also be ecological. Environmental journalist Richard Manning makes the case that, for centuries, life in the grasslands was "nomadic, uncivilized, and therefore hostile to literature," which makes the writing of these absentee authors more, not less credible (195–96).

The problem, however, is that this mobile notion of home, though ecologically informed, does little to counter the culturally informed notions of the grasslands as a place to move through and beyond. In addition, when teaching these regional authors, we often overlook the psychological impact of the destruction of the native grasslands (which several of them witnessed firsthand). Garland, for instance, who was raised on the Iowa grasslands in the late nineteenth century, later expressed shock and pain and guilt over their sudden and seemingly irreversible loss: "The prairies are gone. I held one of the ripping, snarling,

breaking plows that rolled the hazel bushes and wild sunflowers under
. . . I saw the wild fowl scatter and turn aside; I saw the black sod burst
into gold and lavender harvests of wheat and corn—and so there comes
into my reminiscence an unmistakable note of sadness" (3). What role
did this immense ecological loss, this "sadness," play in the decision by
grasslands authors, like Garland, to leave the region?

A big role, if the attitudes of my students are any indication. When
I ask why so many of them want to leave this place, one of the most
frequent responses is a lack of natural beauty and diversity. As a teacher, I
could (and often do) challenge this attitude—there is still natural beauty
and diversity to be found here. But what I suspect I'm facing in such
statements is not primarily a lack of knowledge but rather a profound,
largely unacknowledged pain that comes from residing in a place that
has suffered unimaginable loss of life. I suspect this because I feel it in
myself—the gnawing despair, the hopelessness. Where to go with those
feelings, how to focus them, is one of my biggest struggles as a teacher
and writer of place. One option, I tell my students, is to go to the
literature, to locate and claim a community of writers who, unlike some
in the past, have overcome that pain to commit themselves to living in
and restoring this bioregion—writers such as O'Brien and Hasselstrom
and Swander, whose numbers are increasing every year.

Another, more essential option is to go to the land itself and allow
it to make the appeal directly. The initial result, as I've both witnessed
and experienced, can be even more pain. A few years ago, during a
class visit to a preserve in the Loess Hills, an official with the Nature
Conservancy explained that when the organization first purchased the
property, the previous owner had already sold one of its massive hills to
be used as landfill at a nearby construction site. This was a binding legal

agreement, but through negotiations, the Conservancy was allowed to choose which of the hills would be sacrificed. After conducting an ecological survey of the hills, the Conservancy reluctantly made its selection. Shortly thereafter, the hill and all it contained was completely gone, nothing left but a "great blank" that contrasted sharply with the surrounding healthy prairie.

"My God," a woman responded, her voice trembling, "I can't believe it. It's like *Sophie's Choice*."

This student, Susan, was referring to the movie (based on the William Styron novel) in which a woman is forced to choose which of her two children will be sent to a Nazi death camp. On the one hand, this is a big conceptual stretch, one that she and the rest of us might have dismissed. On the other hand, her emotional response after hearing that story, in that place, made real and personal what had previously been an abstraction, confined and softened by the walls of the classroom—the central grasslands as a site of human and ecological genocide. She later wrote a moving personal essay on the subject, integrating this story and additional research, and has since regularly volunteered on prairie restoration projects. In the face of profound loss, she has embraced the mandate to witness, to be vigilant, and to work for positive change. And to stay where her talents are most needed.

Like many, I am haunted by my failures as a teacher—the students I couldn't reach, the misinformation I've accidentally disseminated, the times I've let anger or despair get the best of me in the classroom. Sometimes I feel that my teaching, like the prairies, is in a constant state of recovery. Sometimes I still feel like an idiot, aimlessly wandering from one class, one committee meeting, one semester to another. Sometimes I just feel like running away. It's at times like these that I come back around to Susan and that day on the prairie, and am reminded that successful teaching about place is not always about complex pedagogies or structured service-learning projects or proper text selection. It can also be about creating quiet opportunities—for the students and for me—to learn from the land itself. And what I've learned is that the prairies here, though diminished, still have the power (as they did for Albert Pike more than a century ago) to throw us back upon ourselves. Not just back upon our knowledge or ignorance about bioregional history and literature and culture and ecology, but all the way back to the foundations of our ethical life. The foundations of our humanity.

This is the gift of the prairies *as they are*, and it is a gracious one. The place that none can afford to leave behind.

COURSE READINGS
for "Grasslands Autobiography"

Bair, Julene. *One Degree West: Reflections of a Plainsdaughter.*
Butala, Sharon. *Perfection of the Morning.*
Garland, Hamlin. *Son of the Middle Border.*
Gruchow, Paul. *Journal of a Prairie Year.*
Hasselstrom, Linda. *Feels Like Far: A Rancher's Life on the Great Plains.*
Heat-Moon, William Least. *PrairyErth.*
Manning, Richard. *Grassland: The History, Biology, Politics, and Promise of the American Prairie.*
Momaday, N. Scott. *The Names.*
Norris, Kathleen. *Dakota: A Spiritual Geography.*
O'Brien, Dan. *The Rites of Autumn: A Falconer's Journey Across the American West.*
Price, John. *Not Just Any Land: A Personal and Literary Journey Into the American Grasslands.*
Sandoz, Mari. *Old Jules.*
Swander, Mary. *Out of This World.*
Young Bear, Ray A. *Black Eagle Child: The Facepaint Narratives.*

WORKS CITED

Anderson, Chris. "Teaching Lewis Thomas." In *Literary Nonfiction: Theory, Criticism, Pedagogy*, edited by Chris Anderson. Carbondale: Southern Illinois University Press, 1989.

Cooper, James Fenimore. *The Prairie*. 1827. Reprint, New York: Penguin, 1967.

Dickens, Charles. *American Notes and Pictures from Italy*. 1842, 1846. Reprint, Oxford: Oxford University Press, 1987.

Erdrich, Louise. "Big Grass (Northern Tallgrass Prairie, North Dakota)." In *Heart of the Land: Essays on Last Great Places*, edited by Joseph Barbato and Lisa Weinerman. New York: Vintage, 1996.

Flores, Dan. "A Long Love Affair with an Uncommon Country: Environmental History and the Great Plains." In *Prairie Conservation: Preserving North America's Most Endangered Ecosystem*, edited by Fred B. Samson and Fritz L. Knopf. Washington, D.C.: Island Press, 1996.

Garland, Hamlin. *Prairie Songs*. Chicago: Stone and Kimball, 1893.

Gayton, Don. *Landscapes of the Interior: Re-explorations of Nature and the Human Spirit*. Gabriola Island, B.C.: New Society, 1996.

Jackson, Wes. *Becoming Native to This Place*. Lexington: University of Kentucky Press, 1994.

Leopold, Aldo. *A Sand County Almanac*. 1949; New York: Ballantine, 1984.

Manning, Richard. *Grassland: The History, Biology, Politics, and Promise of the American Prairie*. New York: Viking, 1995.

Ryden, Kent C. *Mapping the Invisible Landscape: Folklore, Writing, and the Sense of Place*. Iowa City: University of Iowa Press, 1993.

Worster, Donald. *The Wealth of Nature: Environmental History and the Ecological Imagination*. Oxford: Oxford University Press, 1993.

CHAPTER 11

The Bayou and the Ship Channel
Finding Place and Building Community in Houston, Texas

TERRELL DIXON AND LISA SLAPPEY

Most Americans view our cities as having a sort of built-in placelessness. Cities are mainly freeways, and malls, and chain restaurants. The real Places, the ones that matter, are the Grand Canyon, and Old Faithful, and Rocky Mountain National Park. Once we allow for differences in weather, one American city is pretty much like another.

Our students share this general sense that place is somewhere outside the city. To counteract this viewpoint, to get the students thinking about cities in a different way, we each have chosen to focus our "Literature and the Environment" classes on the city where we teach: Houston. At first, it sounds like an overwhelming challenge: how can we make the epitome of the placeless city a place that lives for them? The solution is simple: get them out in it.

What follows is a distillation of our experiences in teaching urban nature in Houston. Lisa works with undergraduate students, mostly non-English majors, at Rice University; Terrell works with graduate students in creative writing and literature at the University of Houston. We share the desire to help our students discover the nature of place in the city. Taking Rice students to the Ship Channel and University of Houston students to Buffalo Bayou makes Houston both our subject and our classroom.

THE SHIP CHANNEL

When I ask students where they imagine living after college, few say Houston. As a small private university within Houston's metropolis, Rice can often seem dissociated, both physically and culturally, from the surrounding city. The school boasts a strong residential-college system, and even commuting students need not stray far from campus. Students rarely see this city as a desirable living place, nor do they realize how many Rice graduates either stay in or return to Houston to work in our medical center, our petrochemical industry, or our shiny downtown buildings. Many will make their fortunes and raise their families right here. To begin to instill in them an awareness of Houston as a wonderfully complex place, we go outside for a look at the city that they call home for at least four years.

What they see is not always pretty. A trip to the Ship Channel may provide the most revealing, if the least aesthetically pleasing, vision of Houston. Though we may also visit Buffalo Bayou, Brazos Bend State Park, or the Galleria, the Ship Channel takes us to the source of so much of the filthy lucre available here. Many of the social, economic, and environmental challenges facing Houston are evident in this segment of the waterway that connects Buffalo Bayou to the Gulf of Mexico. The place itself is a history lesson. In 1836, Sam Houston's forces defeated Santa Anna's Mexican Army at the Battle of San Jacinto, opening the way for an influx of American settlers into the new Republic of Texas. Most of those early arrivals to the city founded by the Allen brothers and bearing Houston's name came by this water route. The outcome of that 1836 contest of international wills set the stage for Houston's emergence as not merely a Southern cotton exchange or oil refining boomtown but as a global commerce headquarters as well.

Houston houses the nation's second-busiest port and the world's second-largest petrochemical complex. Proud of those designations, the city offers a free public relations boat tour of the Houston Ship Channel. Taking my students on the ninety-minute tour aboard the M/V *Sam Houston* to meet our petrochemical neighbors requires that we make reservations, bring identification, and clear Homeland Security. Bureaucracy aside, everyone associated with the Ship Channel has always welcomed us, though I warn the class not to expect the "unforgettably spectacular waterborne adventure" promised on the Port of Houston's official Web site.[1] We journey from our little urban oasis across the 610 Ship Channel Bridge and onto Clinton Drive, passing parking lots full of shiny imported cars on one side and ramshackle housing on the other. Ship Channel neighborhoods lack the Rice area's affluence. Those who can afford to live elsewhere do. Environmental justice begins to take human form.

Students ask why Houston would market this view of itself to tourists, and yet the ship's register is full of visitors. This area interests our guests more than it interests the locals. Over the years, only two of my local students had taken the boat tour before our field trips. Virtually all admit to a basic lack of awareness of the Ship Channel's functions. Some are surprised to learn that we are a port city; certainly they are oblivious to the port's past and present connections to Rice University. Marie Phelps McAshan describes in romantic terms the 1874 port "dream" of Houston as a cotton exporter. This economic and engineering feat required "the narrow tortuous channel choked with logs, infested with alligators, and full of all kinds of debris" (100–101) to be dredged to a depth of twenty-five feet; it succeeded because "these men of great vision and indomitable energy persevered" (101).[2] The project was completed in 1914. The leaders in the later stages included Baldwin Rice, Houston mayor and nephew of Rice University's founder, and Jesse H. Jones, namesake of Rice's Graduate School of Management.

Overwhelmed by the Ship Channel, the noxious smells, noise, smoke, the huge refineries, the filth in the water and along the banks, my students begin to understand Houston's prominence in the global petrochemical economy. Oil and gas politics underpin many national strategies, and the Port of Houston, at once dangerous, powerful, and vulnerable, becomes a significant place. Recognizing the port as our neighbor makes us notice when, for example, refineries explode, as they do, faithfully and conveniently, at least once a semester.

For those who live and die by the petrochemical industry, there is nothing convenient about the constant threats and occasional eruptions that serve as student lessons. Once students understand that we all face environmental perils daily, they ask how individuals can initiate change when corporations seem so powerful. Sandra Steingraber reminds us in *Living Downstream* that "with the right to know comes the duty to inquire" (xxii) and, ultimately, "the obligation to act" (117).[3] This is not easy, since closing ranks is industry's typical response to disaster.

On February 6, 2006, explosions at Akzo Nobel Chemicals in Deer Park sent one employee to the burn unit. Officials shut down nearby roads while attempting to determine the danger level; instead of calling the city's firefighters, the plant's in-house fire crew handled the situation. When one student investigated this incident for her midterm project, neither Akzo Nobel nor OSHA provided any information relating to the fire, so she exercised her right to know and fulfilled her duty to inquire by filing for disclosure under the Freedom of Information Act.

On the boat, we wonder that the industries lining the shore do not have more frequent mishaps than those making the evening news. Perhaps nothing less than spectacular tragedy will force our attention, however briefly, to what occurs along the Ship Channel. How many of us know or care about the chemical processes and political maneuvers that keep the Ship Channel—and therefore much of Houston— generating growth and income? Who pays attention to the daily Air Quality Index, monitors particulate matter, or worries about toxins in the water? Should a hurricane wind its way up the Ship Channel, what sort of unnatural disaster might ensue? In November 2004, *Texas Monthly* magazine's executive editor S. C. Gwynne described the Ship Channel as a "prime terrorist target" because "it is both ground zero for the nation's petrochemical industry and home to unfathomably large quantities of the deadliest, most combustible, disease-causing, lung-exploding, chromosome-annihilating, and metal-dissolving substances known to man"[4]; yet until the Bush administration planned in 2006 to award management of several U.S. ports (Houston was not among them) to Dubai Ports World, how many citizens recognized the port systems' vulnerability to any form of terrorism?

Despite its activity level, there is a depressing, even desolate quality to the Ship Channel. Advocates of retrofitting industrial plants could build their cases here: much of the equipment seems old, rusted, and run-down; some structures, such as the Houston Lighting and Power building, are simply abandoned and falling apart. Add a few dark clouds, and the view becomes downright ominous with silhouetted spires belching toxicity into the air. Here where the hot, hard, dirty work takes place, the effect of heavy industry cannot be disguised. Unless we are doing that work, unless we live in the communities where such work takes place, it is easy enough to disregard it. On the boat's return trip, we see Houston's glassy, expensive downtown rising in the background.

The Ship Channel provides olfactory as well as visual stimuli. One class suggests renaming the boat trip "The Tour of Smells." Author Bill Minutaglio recounts a "joke" popular in Texas City in the 1940s: *"What brought you here?" "Shit, that's an easy answer. It's the stink. The stink of money."*[5] The Port Authority takes pride in "generating more than $10 billion in business each year and supporting nearly 300,000 jobs."[6] In 2005, ExxonMobil alone, whose Baytown, Texas, plant is the largest refinery in the United States, posted a staggering $36.13 billion profit.[7]

Although it is easy to revile petrochemical giants for their avarice, they are not the port's only financial beneficiaries. According to *The Port of Houston Magazine,* each year more than 6,000 vessels carrying 15 million tons of containers loaded with consumer products pass through the facility. They are destined not only or not even primarily for metropolitan Houston, but for all of the pretty places in our nation's interior. The port's motto, "The Port Delivers the Goods," could not be more appropriate. Many of us sell these goods; all of us use them. The boat tour reminds us that everyone is implicated in the activities of the Port of Houston and the Houston Ship Channel. The stench, which is often what students recall long after the tour, clings to all of us, no matter how far we distance ourselves from the Ship Channel.

Environmental lawyer Jim Blackburn lectures in the Department of Civil and Environmental Engineering at Rice University and opposes the massive new Bayport Container Terminal. In his beautiful text, *The*

Book of Texas Bays, he urges us to consider the personal implications at the intersection of ecology and economy:

> We Texans are on a path whereby the full costs of our projects—our water supply, our shipping channels, our water pollutants—are not being paid today. . . . Our most pressing need is to find a different way of thinking about economics and ecology, one that is serious about protecting life on our living planet, one that provides for an accounting of the full cost of our current activities.[8]

Blackburn's work, including a federal lawsuit to stop the Bayport project, reveals the highly politicized issues surrounding the Houston Ship Channel. Although the suit was rejected, at least the Port Authority's Web site now includes a nod toward environmental consciousness. At 45 feet deep and 530 feet wide and crowded with hundreds of industrial facilities, today's Ship Channel bears little resemblance to that wildly lush, narrow channel through which early settlers journeyed up Buffalo Bayou to Houston. It is a fine place to account for the costs not just of our projects but of our way of life.

BUFFALO BAYOU

As Houston has grown, its bayous have suffered. Some very early city plans followed the flow of the bayous, but after that our bayous have more often than not been degraded and ignored. Some remarkably sinuous streams that once threaded the landscape of the coastal plain have been "channelized" (that is, in a process as ugly as the word itself, reshaped into a straight line and then cemented). Others have been made unappealing ditches, dumping grounds for trash, homes to nests of plastic bags, shopping carts, abandoned cars, and the occasional murder victim. Though we citizens usually choose not to really see our bayous, we are sometimes inadvertently implicated in a drive-by bayou sighting. We glimpse part of a bayou, usually from the corner of our eye, from a car speeding down a crowded freeway. Such happenings have inspired us, not to stop, look, and walk but to keep on going. In a revealing synecdoche that signifies how we Houstonians have treated our home landscape, the nickname Space City has mostly supplanted the original, place-based Bayou City. We have chosen to look away.

The problem has been, I think, that the bayous were not part of the aesthetic vocabulary of most Americans. Without peaks and rushing streams, it took us some time to learn how to read their beauty.

Houstonians were prodded into this recognition process by two disparate things. One was popular culture on the national level; the other was the work of Richard Florida, a Carnegie Mellon professor who is a specialist on urban economies. The popular culture provocation was the late-night television world of Jay Leno and David Letterman. Houston has always been an exceedingly image-conscious city, mindful of its status relative to other cities and determined to advance its place in the great, fluid hierarchy of city reputations.

Thus what transpired during the presidential election of 2000 had a huge impact: Leno and Letterman made fun of Houston! The jokes began with a George W. Bush campaign promise that he "would do for the country what he did for Texas." The talk show hosts tied that phrase to then current news stories about Houston's number one ranking in air pollution, Houston's traffic gridlock, etc. For Houston power brokers, the ridicule translated into a loss of face and, even worse, a potential loss of business. At last change had to be considered.

My seminar students at the University of Houston come from all parts of the country, and they begin the seminar with the Jay Leno humorous view of Houston. No need to teach it. They have already discovered poor public transportation, high heat and humidity. It is a big, hot city, and they mostly try not to get out in it, except in an air-conditioned automobile.

The change in the city's attitude toward bayous also stems from the theories of Richard Florida. Sometimes we read his work; most often I summarize it. Florida has developed—with appropriate data, charts, etc.—a profile of what he designates the "creative class."[9] This class of young people, he says, now drives the economic engines of the cities where they *choose* to live. "Choose" is the keyword here; Florida believes that those who make up the creative class seek more than a well-paying job. They take the ability to earn a good income as a given. What they look for, instead, is something that Florida describes as "quality of place." Quality of place for these young people has many different components, but foremost among them are such "amenities" as diversity. To them, diversity means everything from an ethnically mixed population, to varied kinds of restaurants, to active gay and lesbian communities, to an atmosphere where current artistic creation flourishes (not the big museums that emphasize the work from other times and places). It also encompasses opportunities for outdoor activity, and natural places conducive to outdoor recreation are a high priority in this group's evaluation.

Clearly, high air pollution and traffic gridlock will not attract this group. And, just as clearly, the landscape feature in Houston that has the most potential is the bayou system. It seemed no accident that Richard Florida's work began to appear on the Web site for a relatively new organization, one whose membership overlapped greatly with that of the

Greater Houston Partnership (formerly called the Houston Chamber of Commerce). This organization calls itself the Quality of Life Coalition, and, since a fifty-page essay written by Florida is part of the Web site, the echo of his phrase seems deliberate. The negative push of Leno and Letterman and the positive pull of Florida's argument combined to have a powerful impact. At last, a substantial change in Houston's long-term complacency about ongoing landscape degradation seems to be happening.

My graduate students' notion of what a city can and should be also changes as they learn Florida's views. Their sense of Houston, however, changes most with the group field study segment of the seminar. This usually involves two visits, the first to a cement-bottomed bayou not far from campus, complete with high-wire lines and rusted pipes over the bayou. This serves as my example of a "bad"—that is, "unrestored"— Houston bayou. During one recent early-spring visit, however, there were great blue herons in the water, and a sharp-eyed student spotted huckleberries growing along the bank, enough for two very tasty pies— reminders that it is hard to completely eradicate nature. We may have to work to see it, but it is there, even in the heavily industrialized city.

This changing sense of what was best for Houston crystallized in "Buffalo Bayou and Beyond: Visions, Strategies, and Actions for the Twenty-first Century." A long title for a huge plan. The project is, both in scope and in finances, typical of Houston at its most ambitious: the redevelopment of some ten miles of downtown bayou, to be completed over a twenty-year period, at a projected expense of $800 million.

The plan also was a stunningly attractive departure from business as usual in Houston—a needed fulcrum for how the city treats nature. It is by no means the first plan to beautify this bayou; there were at least twenty such proposals, of varying scopes, during the twentieth century. None of those earlier plans had either the range or the backing of this one, and most of them got nowhere. This was good news. It also has made it more uplifting to use the city as environmental text.

With the new plan, it suddenly seemed as if Houston was ready to move into a new growth stage, a maturity that honored, rather than degraded, the natural world. I went to the announcement, noted that several people talked with pride about "greening Houston," and left with only one nagging question. Some dignitary mentioned an aquarium, but I couldn't find it on my expensive, detailed take-home rendition of the plan. I wondered: do we need an aquarium there, right on the banks of the bayou? Even if the aquarium is presumed to have educational value, doesn't that make either one or the other of the two bodies of water—either the real bayou or the man-made aquarium—redundant?

I soon found out. On October 20, 2002, less than a month after the big introduction of "Buffalo Bayou and Beyond," the *Houston Chronicle* featured a front-page story praising a new draw for tourism and a boost for the economy: the Downtown Aquarium. The news that morning seemed so far from the elaborate press conference presentations that I had to read it several times. That was because the Downtown Aquarium is a mammoth business venture—a $38 million aquatic-themed entertainment center spread over six acres, featuring the following: a huge seafood restaurant, a 200,000-gallon shark tank with a mini-train running through it, the capacity to hold three thousand people at a time, and a merry-go-round with sea life rider seats near a ninety-foot Ferris wheel. At night, the Ferris wheel glows bright blue neon. This place is, beneath the self-serving designation of itself as an aquarium, a very large and Disney-fied Landry's seafood restaurant, one with a permanent carnival attached and several large fish tanks squeezed into a small space below the restaurant and gift shop. It was the first bayou project after the September "Buffalo Bayou and Beyond" announcement, *and* it was to be on six acres of public land. The land deal was so blatant that the Houston papers—even the alternative one—dodged the underlying issue.

The argument, widely circulated by word of mouth, was that the Ferris wheel and the carnival were needed to draw "more people" downtown. "To thereby build community as we build downtown." A key point was that "not everyone would go to the Wortham Center Opera House across the street."

The students review this decision, the company involved in it and its political connections, and what was written about it. We talk about the culture's need, first, to do away with nature and then to build commercial, profit-generating simulations of it, about what happens when city governments try to outsource the responsibility to provide parks for their citizens. The rationale for the city's aquarium decision, as I suggest to students before we walk this section of the bayou, underestimates both the power of nature and the priorities of the majority of those Houstonians who are working class or middle class. A real park could build a more inclusive community; the city's decision assumed, and thereby strengthened, a socioeconomic- (and often color-) based separation. People from all walks of life desire and deserve parks, both to enjoy the natural world and to enjoy sharing space with someone who is not just exactly like them.

Since the 2002 announcement, Buffalo Bayou has prospered. Two stages of a beautiful and environmentally helpful redevelopment are done (and the two stages do include some areas with diverse residential demographics). Though I still contend that the permanent "seafood" carnival detracts from downtown and trashes the bayou, the overall Buffalo Bayou redevelopment progress has been good. Houston, downtown, on Buffalo Bayou, is making progress. It is green and spacious, with a good view of a mostly beautiful skyline (marred only by the bright neon Ferris wheel) in the distance, good trails, a great dog park, and the chance to see large turtles sunning themselves on bayou logs and many migrating birds in the spring and fall. I urge my students to walk there at least several times during the semester.

There is a certain pleasure and heuristic value, however, in the field study trip when the seminar goes downtown to walk the bayou. The discussions and readings work well, but the main field study segment of the class helps multiply the impact of the students' library and classroom work. However contradictory the term "urban nature walk" may sound, it works. It remains the best teaching tool that I have.

We start at Sabine Street and stroll down toward the aquarium. This juxtaposition of Buffalo Bayou and the aquarium entertainment complex built on its banks does have one redeeming virtue. It creates a serendipitous site for students to observe and learn about contemporary American culture and urban nature's contested place in it. On our way, we see turtles and birds—herons as well as egrets. Often, we see alligators. Students have already read a funny short story by Donald Barthelme, a legendary teacher in our graduate program. Barthelme's "Return" creates a contemporary urban everyman who returns to Houston and tries to connect with the city through nature, but who finally settles for a simulation of Southern nature, a nine-hundred-foot steel azalea.[10]

When we get to the "aquarium," we note the exhibition room that displays stuffed or glassed-in versions of the local wildlife that we have just seen. The name over that door says "Louisiana Swamp." We pass it, and then head back up Buffalo Bayou. Students sometimes ask, "Where do they keep the steel azalea?"

Most of these seminar students will leave Houston. Only one or two from a class of fifteen will teach and write here. But whether they settle in small college towns or teach in urban universities across the country, they do leave with a sense of one city's efforts to re-green itself, and of its missteps and its successes. The degree of commitment will vary, but they are all to some degree part of a larger community that has started to recognize the importance of urban nature. They can take with them the knowledge that more than 80 percent of their students will live in cities, that city wilds and walks within these wilds are crucial if we are to expand America's environmental consciousness.

CONCLUSION

The Ship Channel and the Bayou embody the two central attitudes that the city of Houston has had toward the natural world—the Ship Channel expresses Houston's past, build-and-develop-the-hell-out-of-this-place-and-damn-the-consequences attitudes. It may someday get cleaned up a bit.

Upstream and uptown, there is Buffalo Bayou, increasingly seen as an "amenity," a way to attract the labor force, those imaginative city builders of the future—beautiful in itself and becoming developed, mostly, in ways that will enhance that beauty and the city's enjoyment of it. Houston's future, we hope, will include the beautification of other bayous. But certainly this is progress. Houston is not now and never will be Portland, or Boulder, or San Francisco. It is, however, now a city where community has begun to coalesce and grow around the idea of finding the place—the bayous, the trees, the neighborhoods—once hidden beneath our decades of determined growth into placeless-ness.

Cities, wild places before they became urban centers, can be made green again, and our students at the University of Houston and at Rice can someday support such a process wherever they live. They learn that they want a city with jobs and homes, *and* places that can embody what Terry Tempest Williams calls "the open space of democracy."[11] They seek places where citizens of all kinds can go to renew their human connections and discover a larger sense of community, their ties to the green world that undergirds and supports the life of the city.

NOTES

1. http://www.portofhouston.com/samhou/samhou.html (accessed September 3, 2006).

2. Marie Phelps McAshan, *A Houston Legacy: On the Corner of Main and Texas* (Houston: Hutchins House, 1985).

3. Sandra Steingraber, *Living Downstream* (New York: Vintage Books, 1997).

4. S. C. Gwynne, "Attack Here." *Texas Monthly,* November 2004. Posted on the Houston Architecture Info Forum, November 1, 2004. <http://www.houstonarchitecture.info/haif/lofiversion/index.php/t618.html> (accessed September 4, 2006).

5. Bill Minutaglio, *City on Fire: The Explosion That Devastated a Texas Town and Ignited a Historic Legal Battle* (New York: Perennial, 2004), 7.

6. *The Port of Houston Magazine,* January/February 2006, 2.

7. "Exxon Mobil Posts Largest Annual Profit for U.S. Company," *New York Times,* January 31, 2006. <http://www.nytimes.com/2006/01/30/business/30cnd-exxon.html?ex=1296277200&en=8ec83a7f4025b22b&ei=5088&partner=rssnyt&emc=rss> (accessed September 2, 2006).

8. Jim Blackburn. *The Book of Texas Bays* (College Station: Texas A&M University Press, 2004), 56–57.

9. Richard Florida's thesis is developed in *The Rise of the Creative Class* (New York: Basic Books, 2002). The Houston Quality of Life Coalition Web page presents (under the heading "Additional Resources") a fifty-page essay that includes many but not all of his ideas: "Competing in the Age of Talent: Quality of Place and the New Economy."

10. Donald Barthelme, "Return," in *The Teachings of Don B* (New York: Turtle Bay Books, 1992).

11. Terry Tempest Williams, *The Open Space of Democracy* (Barrington, Mass.: Orion Society, 2004).

CHAPTER 12

Rediscovering Indian Creek
Imagining Community on the Snake River Plain

ROCHELLE JOHNSON

> "People of every place and time deserve a history."
> —Joseph Amato, *Rethinking Home*

Freshmen who arrive in the city of Caldwell to begin their first semester at Albertson College of Idaho most likely take Exit 129 off of Interstate 84, which runs through southern Idaho. Exit 129 leads them to Twenty-first Street—the bane of my college administration's existence. In all seriousness, it's a problem for recruitment, for retention, and for campus morale. Before reaching the college, Twenty-first Street enters a half-mile business development called Farm City. This poorly tended strip is about as depressing as western America can get: it features a small, fenced-in mud pool that serves as home to about a hundred crowded cows; piles of rocks, railroad ties, and concrete machines; no vegetation whatsoever (well, except cheatgrass); and—this is the worst—a prominent billboard with large letters boasting disparaging comments about the college, the "liberal" media, and the downfall of free enterprise as we know it. "Welcome to Farm City!"

In fact, the entire city of Caldwell is pretty depressing. When I arrived here in 1999, I found a once-busy rural western town largely deserted and depressed economically. Ten blocks or so from the college, the buildings of the downtown area displayed far too many "For Rent" signs, and the sidewalks were deserted. The shops that remained—the pawnshop, the gun shop, the area thrift store, and the independently owned department store with shelves of dusty items—were struggling. One day a few months after my move here, the only car wash in town literally collapsed. (Luckily, no one was inside it at the time.) Caldwell needed help, and because the condition of the town discouraged so many of our prospective students from attending the college, my place of employment needed help, too. And to be honest, I needed help. To my New England family's utter surprise, I had decided to try to make this place home. After my initial shock, I committed myself to the place because of its nearby mountains, its sagebrush-covered deserts, and its grand skies. But developing a sense of place in Caldwell was proving difficult.

In the midst of my disenchantment, I was surprised to learn that the car wash hadn't merely collapsed; rather, it had fallen into a creek. Creek? What creek? It was then that I learned that much of downtown Caldwell rests on concrete and rebar slabs that were used several decades ago to cover up an unsightly, filthy waterway: Indian Creek, which flows some fifty-five miles through the Western Snake River Plain of southern Idaho and meets the Boise River just west of town. Once the very reason for human settlement here, the creek had been covered during the mid-twentieth century by streets, parking lots, and buildings. Most area residents were ignorant of the creek's existence—they were too young to have known it, too old to care about it anymore, or new in town, like me. On a walk one day near campus, I realized that Indian Creek also runs under Twenty-first Street in Farm City. While driving down that street, incoming students just might catch a glimpse of the creek between the railroad ties and the rock pile.

By 2001, city officials began to wonder if raising awareness of the creek might help address their economic problems. The plight of the car wash actually helped them toward their dreams, since part of the creek was now showing itself. Many citizens were against this plan, but soon city leaders were suggesting uncovering—or "daylighting"—Indian Creek. Feeling directionless and economically vulnerable, city leaders saw Indian Creek as a potential element of an attractive downtown— one that regularly featured people walking, shopping, and visiting. Their dreams involved making the creek the town centerpiece; they imagined walking paths and benches along the waterway, and nearby coffee shops, businesses, and restaurants. They dreamed large: the individual who had first suggested uncovering the creek went so far as to envision a water wheel in a small park right downtown. As a member of Albertson College of Idaho's environmental studies program, I was invited to participate in what came to be known as Caldwell's "downtown revitalization project."

While I felt the satisfaction of being involved in shaping my place by participating in these efforts, I also felt that some things were missing from our discussions. For example, though we spoke of making the creek more "natural," the term lacked much meaning. What had the creek once been? What was its history? Furthermore, we were largely ignoring the many people who were adamantly opposed to uncovering the creek. Many citizens remembered a filthy, smelly creek filled with rats, carcasses from nearby meat-processing plants, and, rumor had it, typhoid. Furthermore, over the years several children had drowned in the creek's rushing waters. Why daylight a dangerous and dirty creek?

Increasingly, recovering the story of Indian Creek seemed to me to be crucial to helping people imagine its physical recovery. As plans for the city progressed, I felt compelled to help build a community-wide sense of ecological responsibility for this neglected creek. I imagined a way to do so through my teaching. I designed two sections of an advanced composition course that would work together to write a book about the creek. We would focus the book on the creek's natural history, and when the book was complete, I would distribute it throughout the community. I believed that the story the students would uncover could lead them to feel responsible for the long-term health of the creek. While I didn't know much of the creek's story myself, I also trusted that the rich natural history we would discover could help inform those citizens who opposed the change. Ultimately, I hoped that the project might also help my students feel connected to a community they currently had no interest in. And I secretly hoped that I might come to feel closer to this place, too.

During the spring semester of 2003, the thirty-six members of my two sections of the advanced writing course worked collaboratively to write the history of Indian Creek so that community members and city planners might bear in mind its long past as they envisioned its future. The final result was *Rediscovering Indian Creek: The Story of Our Region,* a fifty-nine-page illustrated volume that was printed locally, and in which the students took much pride. When the course began, however, I quickly discovered that most of my students did not care very much at all for Caldwell—or for the Indian Creek sub-watershed. In fact, most of them announced on the first day of class that they felt little connection to the area in which they lived: "I just go to school here," they said. But we forged ahead. Following my suggestions of some broad topics, the students conducted their research through books and interviews, determined the specific contents of the volume, and began drafting a manuscript.

Right away, the students saw that the stakes were high in this course. First, they would be writing for a "real" audience. Not only would they actually see their efforts in print, but someone other than their professor would be reading their work. Second, they would have to rely quite seriously on each other. While the students would perform their own research and write drafts individually, later they would combine their own contributions with other students' essays on similar topics. They saw that they would need to agree on criteria as they exercised literary judgment, and suddenly it seemed that seeking out other models of natural history writing was the most important way to spend their spare time. Third, they realized that in order to acquire the history of Indian Creek, they would need to work with people outside of the comfortable campus community. They would collaborate not only with each other but also with city leaders, the Caldwell School District, the National Park Service, the Army Corps of Engineers, and, most important, local residents and area experts on local history, geology, ecology, native species, and stream restoration. They would need to make contacts, ask questions, and reconstruct the story of this place. Finally, the nature of place-based writing would require them to work with material outside of their areas of comfort.

Indeed, I later delighted in watching the geology major explain the formation of Idaho's river valleys to two English majors and overhearing a philosophy major explain to five biology students the ethical implications of displacing area businesses as the creek was uncovered. Several students met with an alumnus from the college who recalled testing the waters of Indian Creek for a biology class in the 1960s. What the tests had revealed then was typhoid. This information compelled that particular group to explore the current levels of toxins, sediments, and nutrients in Indian Creek. Working with a faculty member from the biology department, they tested water, measured sediment load, and taught themselves and each other how to interpret their data.

As they conducted their research on various subjects, the students uncovered a story that compelled their attention—and that later compelled the attention of even skeptical readers. The story of Indian Creek had been one of vibrant life, diversity, and, later, accommodation. To emphasize the fact that this region had been a thriving natural community for many thousands of years prior to any human presence, the students began the book with sections on the geologic origins and native plants and animals of the area. Acknowledging that the pre-human story of the creek was far longer than the human one, they devoted fully half of the book to the long period before human settlement. Turning then to settlement, they researched the Northern Paiute and Shoshone-Bannock peoples, Euro-American pioneering settlers, the non-native species that accompanied Euro-American settlement, the changes brought about by the railroad, and the eventual appearance of towns along the creek: Kuna, Nampa, and Caldwell.

The history of Indian Creek helped explain to my students the depressed nature of our town. For the better part of the past century, the creek had been used as a dumping ground for nearby industry, co-opted as a route for part of the major irrigation system that serves the valley, and covered over by streets, sidewalks, buildings, and parking lots in downtown Caldwell. In earlier times, before legislation forbade such things, businesses and livery stables threw their trash in the creek at the end of each day, and the powerful waters carried animal carcasses and rats through town. Although residents complained about the unsanitary condition of the creek's waters, they could not imagine another way of living in relation to it. It made sense, then, that city officials had undertaken the monumental task of covering up the creek, citing its stench and the need for improved traffic flow as reasons for their decision. Because they could not imagine a story in which the creek was honored for the life it had brought to the region, they did their best to erase its presence from their landscape.

The students were disturbed by what they saw as this ultimate expression of the commodification of nature: people, having ruined what had been healthy and thriving for millennia, tried to hide their mistakes by literally covering them up. This aspect of the story, however, helped the students understand the local skepticism about the creek and about Caldwell—a skepticism that many of them had shared. What they found was that the decades following the creek's concrete burial brought with them just what we might expect from imaginations grounded in conquest: the slow death of cultural vibrancy and of a unique local identity. These decades saw the rise of corporate monocultures. Many locally owned businesses closed, the downtown was nearly vacated, and the once-bustling center of Caldwell became quiet, empty, and, frankly, depressing. Meanwhile, on the large boulevard connecting Caldwell to the next town, up came Wal-Mart, Kmart, and Walgreens. The students recognized that the story of Caldwell was also that of the nation: local losses resulting from a national ideology of possession.

However, the students were especially interested in foregrounding this recent history for readers, as it helped to illustrate both the drastic changes to the creek in a very short time and the potential for what might be recovered. While they decided ultimately to omit their anger from their telling of the story, we had several intense class discussions about the connections between the creek's history, our community identity, and their decided distance from Caldwell at the start of the course. Because the place lacked life and beauty, they preferred not to engage with it. The students now saw that such disengagement was not what the creek needed, and they decided to use their book as a way to educate people about the environmental health of our region.

One group of students grew especially interested in the plant life of the creek. Under the leadership of Thaona, a senior biology major, they researched and wrote about the benefits to the creek of native plants, ones that flourish in our particularly dry climate, attract birds, butterflies, and other insects, and allow fish to spawn and thrive. After meeting with the college's botanist, they summarized for readers the processes by which natural wetlands reduce sediment loads and cleanse the water, enabling fish and other life to inhabit the creek. And they addressed community fears about uncovering the creek by reminding readers that federal legislation in the 1970s had introduced requirements forbidding the sort of dumping that had been such a problem in the past for Indian Creek.

All of the students learned a tremendous amount about this particular landscape. Cristina, a senior, led a group of three students on a series of walks along the creek—starting at its headwaters east of the college and ending where Indian Creek meets the Boise River. They found themselves enjoying riparian vegetation much of the way while also having to climb over, through, or around concrete culverts and, as they got to Caldwell, simply walk the streets while wondering precisely where the waterway lay under their feet. They took video clips as they walked, and later combined them into a ten-minute documentary that would be shown at a town hall–style meeting.

Another group spent a leisurely afternoon getting to know a stretch of the creek just a couple of miles east of Caldwell. This particular section of the waterway meanders slowly and widely through a marshy field, bordered by the highway on the north and busy streets on the east, south, and west. But for several hundred meters, the creek runs as it likely did centuries ago. Much to their surprise, they found peace and beauty here, as well as mallards, a great blue heron, and evidence of beaver. These signs of wildlife impelled them to speak with several local ecologists about the animal life that would have once thrived along the length of Indian Creek.

Matt, a junior, grappled with how to convey the irony that he saw one day as he sat by the creek, contemplating its present state. In the distance an American flag, the supposed ultimate sign of love for this land, was waving. In the foreground, however, were several blatant signs of disregard for the landscape: heaps of trash dumped by the creek, the remains of a concrete project piled high *in* the creek, and, most offensive to this Idaho rancher's son, a dead cow decaying in the sunshine. How could he communicate that this creek embodied the long history of disregard for our national landscape? After several conversations with his writing group, he crafted a moving portrait of the creek as just one more powerful piece of evidence in our destructive environmental history.

Another group of students contacted an eighty-five-year-old woman who had lived in Caldwell her whole life. Dessie Cole had grown up in a house right on the creek, and when she married and had a family herself, she still lived close to Indian Creek. One afternoon at her nearby retirement home, several students listened to her stories. She remembered Indian Creek as a place where her father and brothers fished, as a summer swimming delight, and, after it was channelized in concrete, as a rushing, dangerous waterway that she kept secret from her sons as long as she could. While she regretted the course that the creek's history had taken throughout her life, she expressed to the students her love for this place. "Caldwell will always be my home," she told them. They felt moved by her tales, and they began crafting them into paragraphs for their book.

After hearing from Dessie Cole about the historically calm waters of Indian Creek, a handful of students wanted to learn more about why the creek now rushed dangerously straight through town. Working with town leaders, they learned that it had been redirected through straight channels. They then decided to research the natural condition of creeks. Finally, they crafted sections for the book that explained the importance of allowing waterways to meander through the landscape, enabling powerful stream flows to slow down. They drew illustrations to show readers that if the creek had sloping banks again, rather than steep ones, it would not only be safer for children but would also foster plant life.

Clearly, their research took the students in a variety of directions, so we relied on class discussions to ensure a shared vision of what the book should be. Most significant to the students was how they might impress upon readers the fact that Indian Creek had a history that reached back far beyond our memories: for hundreds of thousands of years, the creek had provided sustenance for the varied life-forms of our area. By learning the full natural history of our region, the students reasoned, their readers might come to see that we are all merely temporary occupants of this place. We should feel grateful for our landscape and, they hoped, responsible for it. The story that has lasted for centuries—the story of Indian Creek nourishing this region—might replace the far shorter story of its seeming to be a blight upon this land. The longevity of the creek's presence might also help people feel responsible for its health.

By the end of the semester, the students had produced a draft manuscript. This course did not "end" for me then in the usual way—with students walking out of a classroom at the end of a term. Rather, their departure from the classroom represented a new beginning to my work. During the summer that followed, a student intern and I edited their manuscript, preparing it for eventual submission to a Caldwell printer. We proofread, obtained permissions for images, and coordinated the work of the several translators involved in the project. (The students had wisely asked that the book be made available in both Spanish and English, as many members of our community are literate only in Spanish.) In addition, I wrote grants, raised funds, encouraged the college's development office to do the same, and finally, in the fall of 2004, drove around the valley distributing boxes of the book to area libraries, schools, prisons, and businesses, as well as to hundreds of individuals. Through the generosity of several individual donors, the city of Caldwell, local businesses, and Albertson College of Idaho's environmental studies program, two thousand copies of the book have now been disseminated.

The student writers who produced *Rediscovering Indian Creek* then found that their work truly mattered to the community around them. The book was noticed by area newspapers and at local festivals. It has been used in science classes at both the high school and the college levels, and environmental and philanthropic clubs in the region have met to discuss it. An alternative high school nearby invites students to read from the book as they sit by Indian Creek, and then to journal about their experience after doing so. And just last week, I got a call from a local dentist's office asking for more copies for the magazine racks in the waiting room. The book reached a wider audience yet when at the August 2005 White House Conference on Cooperative Conservation, government and conservation leaders received a compendium of 150 case studies in "cooperative conservation," one of which centers on Caldwell's revitalization project and makes reference to *Rediscovering Indian Creek.*

In terms of the creek itself, amazing progress has occurred. Our town has just completed its purchase of all of the properties along the creek's downtown corridor, and design plans are under way for daylighting the creek throughout that area, including allowing it to move in a more natural, meandering path. If all goes according to plan, the creek will see daylight again before the close of the current decade. Rather than remaining ignored, Indian Creek now takes center stage in this landscape, the very source of Caldwell's identity.

Perhaps because such a course was new to me, it provided me with some incredibly valuable lessons, most of which centered on the challenge of inviting students to work for the larger good of a community. As it turned out, in fact, every single student grew to genuinely care about the Indian Creek sub-watershed and its future. As a mark of their dedication to Indian Creek, they decided early on in their research that they wanted readers to know that they had written this book on behalf of the creek and the community, and not simply because their college course required it, so they named themselves the Indian Creek Writers Collaborative. The students hoped that their work would be seen as the product of a community-based, collaborative effort rather than simply as a product of course requirements. As one of them remarked to me, "I hated Caldwell when I started this course. Now I'm in love with a creek!" I, too, experienced the power of natural history to shape a sense of place. Once we knew the full story of our home landscape, we all felt a deep commitment to it.

Looking back on the work of my students, I am pleased to see that their book has played one small part in a truly community-driven effort to re-envision our occupancy of this place. That our community imagines itself differently in relation to Indian Creek is perhaps best exemplified by the "demonstration site"—a block-long section of the downtown in which the creek has been daylighted, meanderings reintroduced, a small wetlands encouraged, and walking paths constructed along the now gently sloping banks. There's even a small park and a water wheel. Rather than telling a story of distance, avoidance, and shunning, this reshaped section of the creek tells a story of communion, appreciation, and bubbling waters. This portion of downtown Caldwell is a testament to a reimagined life here: one that celebrates and cares for the natural surroundings, if still depending on bulldozers and concrete to convey that story. In addition to these physical changes, Caldwell holds an annual Indian Creek Festival downtown each autumn, and each year on Earth Day a large group gathers to pick up litter on the banks of the creek. We've taken some small steps in the right direction. We're imagining a different story.

Indeed, this place-based course attests to the transformative power of story. Barry Lopez writes in *The Rediscovery of North America* that one path to reimagining our relationship to the land—one of his "sources of hope" (51)—is "literature, which teaches us again and again how to imagine" (53). Stories, he argues, can help us to reimagine our meaning and, eventually, alter our culture's ideological foundations of greed and possessiveness. Lopez's closing pages offer more than a naive hope in the ability of literature to alter consciousness; he also asserts the powers of community, concluding his book with the potent image of all of us turning to one another, locking eyes, and "sensing" through our fixed gazes that profound change is possible. We have, after all, "only our companions . . . to look to" (58). Through the process of producing *Rediscovering Indian Creek,* my students and I experienced the potential of a story to help a community imagine a different future than the one to which it—and we—had been headed, somewhat like automatons, because that well-known path looked familiar. The story of this place startled many of us into new ways of being in the world.

In our days of commuting, relocating frequently, and privileging the self too often at the expense of those around us, the entire notion of community is in question. Through collaborative work with each other, civic leaders, businesspeople, engineers, and other residents of this place, my students and I had an opportunity to dwell on the meaning of community. We found that by recovering the natural history of our place, we could help to ensure "community," which we now understood to be a group of people dedicated to each other and to a purpose: fostering the ecological, emotional, and economic health of a place. While the story of Indian Creek humbled us by illustrating our insignificance in its complex, millennia-long life, we felt empowered by that humility. And we felt the need to share that story.

NOTE

I want to thank Cristina F. Watson, without whom this project could not have been so successful. As the student intern with whom I worked, Cristina created the design for the book, provided beautiful illustrations, and walked more of the length of the creek than any of us in order to capture its variety through photographs. I admire her work tremendously. I also want to thank Laird Christensen for sharing with me his experience in editing *Recovering Pine River*, a student-collaborative work that was the outcome of a course he taught at Alma College. In addition to making copies of his students' book available to me, he shared advice about the logistics of running place-based writing courses.

COURSE READINGS

Asterisked entries are from *Literature and the Environment: A Reader on Nature and Culture*, edited by Lorraine Anderson et al. New York: Longman, 1999.

*Bass, Rick Bass. "On Willow Creek."
*Carson, Rachel. "Of Man and the Stream of Time."
*Dillard, Annie. "Living Like Weasels."
Hacker, Diana. *A Writer's Reference*. 5th ed. Boston: Bedford/St. Martin's, 2003.
*Kittredge, William. "Second Chance at Paradise."
Lawrence, Gale. *A Field Guide to the Familiar: Learning to Observe the Natural World*. Hanover, N.H.: University Press of New England, 1998.
*Lopez, Barry. "Apologia."
Lueders, Edward, ed. *Writing Natural History: Dialogues with Authors*. Salt Lake City: University of Utah Press, 1989.
*Rogers, Pattiann. "Knot."
*Russell, Sharman Apt. "The Physics of Beauty."
*Sanders, Scott Russell. "Buckeye."

Watershed Writing Collective and Laird Evan Christensen. *Recovering Pine River.* WTW, 2000. (Available through Blackboard.)

*Williams, Terry Tempest. "The Clan of One-Breasted Women."

WORKS CITED

Amato, Joseph A. *Rethinking Home: A Case for Writing Local History.* Berkeley: University of California Press, 2002.

Lopez, Barry. *The Rediscovery of North America.* 1990. Reprint, New York: Vintage, 1992.

CHAPTER 13

Gifts and Misgivings in Place

PAUL LINDHOLDT

Teaching environmental studies to rural Westerners differs from teaching it in cities on the coast. To adopt an easy formulation, made by the press during the 2004 elections, one might typify the rural inland as "red" politically and the coast as "blue." Mainstream America may romanticize western values by making the rural West seem simple and bucolic, but teaching here is rarely easy. It can require a fundamental shift to reimagine one's audience. To complicate the educator's task even further, publications for tourists and backpackers idealize rural western landscapes, representing them as commodities for sale but rarely as places to put down roots and learn to get along well with the neighbors.

As someone who teaches classes in literature and environment in rural eastern Washington, I ask my students to grapple with such issues. I ask how "the tragedy of the commons" might reference more than natural resources in the quality of the lives we lead. In an essay by that title, Garrett Hardin wrote about a hypothetical common ground where ranchers graze cattle and ought to share in a wish to sustain the resource. Instead, all ranchers keep adding more cows or cow-calf pairs to their share till the pasture grows depleted and collapses. That is the tragedy of the commons: that greed and privilege often exhaust the common good. Jared Diamond, in his book *Collapse*, explores such patterns of exploitation throughout a number of eras and civilizations. I aim to widen the semantic range of "the tragedy of the commons" to include more-basic elements in the hierarchy of human needs. Safety and space are species of resources, I maintain, less tangible perhaps than browse and grass but substantial all the same. Some Americans, for example—tied to Old West myths of freedom and power—remain bound to weapons and to their right to use them. And words, like weapons, may be granted constitutional rights, despite the potential dangers inherent in such rights.

IN EARSHOT OF WATER

One spring the Spokane River was running high, arrowleaf balsamroot yellowing the shoulders of the road, snowmelt booming twenty thousand cubic feet a second toward its confluence with the Columbia River. My wife pointed out the window of the car I was driving and cried out. A cat that looked like Weezie, her timid tom that had gone missing two days before, was mousing on a hill beside the river. I coasted the decrepit Honda wagon to the gravel shoulder and set the brake. I was quite convinced that she had the wrong cat, but I never got a chance to learn for sure. Out of the car she leapt and began to call and coax. We made a family tableau—her crouching, me akimbo between the car and the open door. A magpie lit on a telephone line overhead just then, and I turned away from the cat.

Conditioned by my past life as a hunter, I briefly considered dropping that magpie from the wire. Magpies, the most voracious avian predators in the West, rob nests and nestlings. They eat crops and ravage gardens. One hunter claimed the birds like to enlarge wounds in cattle and hop inside to devour them from the inside out. A powerful pellet rifle lay in the car; a friend and I had been sighting it in the day before, and I had forgotten to put it away. No one but Karen would witness my little lapse from the law for a good cause. My musing on the bird overhead came to a quick end, though, when a rock slapped into roadside weeds and the magpie flew. Another rock whizzed overhead, this one colliding with a Yield sign some thirty yards away. The clang echoed off the road.

On a stump behind us stood a young David. From one hand dangled a sling, two laces linked by a leather pocket and whirled. What a crude weapon to master and grow so accurate with! I shouted at him; I cursed. He gazed our way, abstracted, as if regarding us for the first time. Then a cloud crossed his brow, and he slid the sling into the bag at his hip. I had made him sore, I saw, maybe made a foe of him. Narrowly he eyed our car as if to memorize it. And then he turned and angled away along an uphill path. Karen crouched behind the open door. The Yield sign was pocked, I saw, where rocks had struck it.

Less angry than anxious, I was trembling and tasting copper. What if he had set out to pelt us? What if my hurled curse or cry had worked him up enough to draw from his fanny pack a pistol he'd held back? My wife, a criminal defense attorney, knows how dangerous some citizens can turn when they get riled up. She is often warning me not to anger others, and not to challenge folks in cars. For myself, I have lived long enough in the rural West to have witnessed or heard tell of memorable lapses of judgment with guns: greedy and illegal hunting out of season, target practice endangering anyone in range, killing for sheer sport. Two off-duty cops in nearby Suncrest, one an officer, the other a new recruit, had gotten drunk and let fly one night with their firearms out the back of the mobile home where one of them lived. A stray slug had entered a house and lodged in a wall near the head of a woman in bed. Their sporting weapons included a strictly illegal automatic.

It's not the firearms alone that disturb me, though plenty of bad gunplay still takes place around the West. It is the disregard of some of my fellow westerners for people, animals, property, and common sense that astounds. Nor could I claim to be free of its symptoms—I who had just been tempted to drop a magpie from the wire. How can so many of us ignore the freedoms and needs of the various other beings that surround us?

Following our encounter with the lace-whirler, the reckless thrower of stones, I brooded for some time on what I had seen. Late into the night I lay, mired in an ancient quandary of rights and liabilities, privileges and guilt. "In dreams begin responsibilities," Delmore Schwartz had written long ago, and I knew what he meant. Freedom and peace necessitate a weighty set of obligations. George Lakoff characterizes such obligations as "responsibility, which is at the heart of liberal/progressive morality." Such morality "begins with empathy, the ability to understand others and feel what they feel" (62). Still today I am learning what motivates my red-state neighbors in blue-state Washington.

THIS GEOGRAPHY

My particular bioregion, the location of the main campus of Eastern Washington University, lies at the edge of three ecotones, three transitional areas between adjacent ecological communities. To the north and the east, the topography rises and gives way to stands of coniferous forests. Mountains loom there, staggering canyons, swift rivers, and enormous lakes named Priest, Pend Oreille, and Coeur d'Alene. To the south ranges the Palouse, one of the planet's most productive wheat- and lentil-growing regions, with its steep rolling hills and deep richness of volcanic loess soil. To the west lie sere, channeled scablands and shrub-steppe (essentially desert), relieved in spots by potholes and seeps. The word "channeled" denotes the remnant paths of prehistoric floods that coursed across volcanic basalt soil and rock, while "shrub-steppe" names the arid scab-rock lowlands and treeless scrub brush threaded by badger burrows, shallow soil, and rocks. This region, encompassing northern Idaho and eastern Washington, has a political identity commonly called the Inland Empire, a place historically dependent for its economic stability on immemorial natural resources—timber, mining, ranching, agriculture, and fishing among Indian tribes.

Extracting those resources into perpetuity is not desirable or possible, for they are finite. The latest culture war, since the spotted owl clash of the 1990s, concerns flagging salmon populations in the Snake and Columbia river system. Dams erected decades ago have come to function less as flood control and irrigation than as ports for barging crops and paper pulp the hundreds of river miles to the sea. Salmon restoration, as mandated by the Endangered Species Act, costs a billion dollars a year amid business as usual. The Indian tribes—Shoshone, Nez Perce, Kalispell, Spokane, and Kootenai, to name a few—enjoy enough legal standing to press for restoration of the fish, but for some reason they continue to hold back. That battle is only one tense aspect of my place. Just as important for my students to learn about are the eroding farm fields, the soil banks failing partly because of chemical inputs and partly because of unsustainable patterns of use that trigger tons of topsoil to blow and flow to parts unknown. Such rapid cultural transformations can teach college students something. This region is relying more on recreation, manufacturing, and tourism than on natural resources. As those same natural resources dwindle, the quality of life erodes, and our dwindling natural resources come to be more precious than jobs or gold.

LIGHTING OUT FOR THE TERRITORIES

One traditional trait of the western provinces has been free reprieve from social obligation, a respite from moral codes. One may think of (and teach) Edward Abbey, who moved from Pennsylvania to New Mexico, studied European anarchy in graduate school, then settled in southern Arizona, where he began to commend radical activism (including "monkey-wrenching" or sabotage) and helped to found Earth First! Abbey's strain of activist environmentalism was essentially anarchical. The literary legend Huckleberry Finn is aiming likewise to "light out for the territories" at the end of the book, to head west and flee the domineering authority of his prim Aunt Polly. Many of my neighbors would appreciate Huck's longings—being, as they are, just as prone as Huck to mobility and itinerancy, just as skittish of obligation and responsibility. Accountability to others and to place—an aspect of the bioregionalism I advocate in class and practice in person—is hard to champion when so many forces militate against it. Inland Northwest residents do not like others telling them how to live, especially if those others signify government, whose proxy as a state employee I unquestionably am. They will not be told what to do.

Their irascibility is an aspect of place. *Winter Range,* a novel by Claire Davis, who lives on the Palouse, centers on such tensions of mutual accountability. When a renegade rancher goes bankrupt one winter and cannot afford to feed his stock, he spurns those who would dispose of his animals humanely, allowing no one to feed them, buy them, or give them care. In a death spiral of a plot, the rancher starves his animals, starves himself, and sets fire to his banker's home, effectively punishing his community for an economy in transition and for his shortsighted failure to manage his own ranch. In *Close Range: Wyoming Stories,* the collection from which "Brokeback Mountain" comes, Annie Proulx explores the lives of cockeyed sodbusters and limp-alongs who are the living equivalents of a middle finger lofted high. "Indigenous" in these parts is as apt to mean "lawless" as "native-born." Vigilante justice holds sway in remoter regions of the rural West, as the Randy Weaver story shows. After selling a sawed-off shotgun to a federal agent, Weaver refused to appear in federal court. Instead he held out, holed up, found his home under siege, his son wounded, wife dead on his cabin floor. He is an aspect of this place.

If renegade survivalists stockpile food against Armageddon, instructors at our nearby Fairchild Air Force base teach survival methods to prepare pilots for the possible downing of their planes. Some academics might feel like crashed travelers, uprooted from family and place by circumstances, plunked down in strange lands while chasing short-term jobs. Sheer survival as priority—the mere endurance of circumstances—can make the higher aims of self-actualization go dormant. People can't fulfill their hierarchies of higher needs when they feel unsafe. How can anyone care enough about a particular place to gain its intimate acquaintance when the personal needs have long gone begging?

Even within its urban concentrations, though, the American West can stifle. How well I recall a pair of African Americans who relocated to the University of Idaho to play football. They enrolled in a writing class I taught. When I tried to do some environmental education, they were skittish about being outdoors. I could guess why. They had heard about the Aryan Nations headquarters, the compound less than an hour away, owned and run by the Reverend Richard Butler, he whose white supremacist ideology had dubbed black folks "mud people." Once the Christian Identity movement came to impinge upon my consciousness of place, mentions of "identity politics" took on a new edge. To teach about my particular place with honesty today means to factor in that Christian Identity.

Once I moved to the inland West to stay, I had to answer to friends in Bellingham and Seattle who wondered why I had chosen to light out for these territories. I said, and I maintain it still today, that the forces of poverty, ignorance, and fundamentalism can best be countered here. If one aims to effect social change, the territories are the place to be.

During my four years as a professor at the University of Idaho, I designed a course titled "Issues in White Supremacy" for the Martin Institute for Peace Studies and Conflict Resolution. Attorney Salmon O. Levinson established the William Edgar Borah Outlawry of War Foundation at the university in 1929. Since 1948, the foundation has sponsored an annual program devoted to understanding war's causes and the conditions needed to establish lasting peace. How ironic it proved when the Aryan Nations set up shop just north of there in the 1980s. Teaching my course in white supremacy helped me to learn as much, certainly, as the students learned from me.

IN THE CLASSROOM

Environmental educators agree on the need to foster sensitivity to place inside and outside the classroom. Students may gain empirical grasp of the fact that we are animals first—mammals who share 98 percent of our genetic material with chimpanzees—and humans only second. In the primary grades particularly, but also in entry-level college classes, students who are allowed to explore and develop connections to nature in their personal lives appear from research to be more apt to thrive as citizens and professionals. Wisdom deriving from experiences, emotions, memories, and personal histories can encourage a virtual reinterpretation of the world. It can cultivate interdisciplinary sophistication and counterweight the illusion of "objective knowledge" that unbalances the scales in so many disciplinary forms of knowledge. On the most intimate levels, our place-based attachments to the past can enlighten and empower our students to shape the future.

I have my students join me on field excursions, but I also ask them to excavate their personal pasts. The excursions typically take place in a national wildlife refuge where my university maintains a field station. We hike and listen, learn some flowers and birds, all of this a preparation for a classroom component on place-based attachments. There is no way I could ever presume to choose a place for my students; I can only show them the ropes and rewards of visiting Turnbull National Wildlife Refuge. To help them own the experience, we spend some time identifying favorite places from their past and present, and then I ask them to compose three separate essays based upon these exercises:

Day 1: Imagination: Try to imagine your special place from the perspective of someone other than yourself. Project yourself, for instance, into the mind of an animal, bird, or tree you have seen; a character, narrator, or author you have read; or a farmer, rancher, or logger you have met. Write freely about your place, using the voice of this person or this entity as you imaginatively conceive that voice to be.

Day 2: Analysis: Analyze your geographical place by separating it into some of its component parts: e.g., air, water, soil, and the variety of lives they maintain. Alternatively, study how the parts of this place fit together. How do they rely upon each other? How do they integrate to make a whole? Now try to narrow your analysis to some particular human impact. For instance, how do internal combustion engines affect your place?

Day 3: Evaluation: Judge the value of your place according to a single principle. For example, what value does your place hold as bird or animal habitat? As a source of beauty? As a site for human recreation? How about ecosystem services, like water filtration?

These exercises get the students thinking about the value to their lives of special places.

Several years ago I was teaching English 308: "Advanced Exposition," a themed class cross-listed with honors. That class entailed field excursions, environmental and social-justice issues, and techniques of civic writing that can effect social change. Two older students arrived with chips on shoulders and pens in hand to record transgressions they were sure they would find within the liberal professoriate. I thought of them recently when I heard some news about an alumni organization that was paying UCLA students to spy on professors. Students were lured, in exchange for providing recordings of allegedly partisan lectures, by the opportunity to earn a hundred dollars a pop for their reports ("U.S. University").

Mine was a small class, fewer than fifteen students, most of them women. It met two days a week, first in a windowless room with clusters of tables. Later we switched to a fishbowl lounge, with windows and doors on three sides. The peer-group work for that class necessitated a certain trust and vulnerability, a relinquishing of insecurities about one's writing and about the personal disclosures the assignments often prompt. In ways we do not fully understand, the environment shapes identities, or so my research has shown. Memories of beloved places, degraded places, and lost causes are common when I set out to assist students in excavating memories of beloved places from childhood. Poet William Wordsworth's legendary "emotion recollected in tranquility" was as much a component of his poetic project as it is for my students excavating their own pasts.

As with many classes at my regional university, this one had a number of nontraditional students. There was Martha, the blue-haired grandmother, active in family and church, aiming to train as a travel writer. There was Fred, newly divorced and crew-cut, retired from the military, who cradled a motorcycle helmet, a state-required accouterment for his "freedom machine," as he named it. There was Nicky, married mother of two, in her late twenties, open-minded but quiet and shy. There was wanda (she did not capitalize her first name or her surname), an African American single mother who had recently been discharged from the navy and sported an eyebrow ring. She was outspoken, savvy, and cocksure. The other members of the class were more traditional students from a variety of majors, most of them nearing the end of their college careers, freethinkers who found the class praxis attractive— the field excursions, the environmental and social-justice studies, and the techniques of civic writing that I had developed as the core of the course.

And then there was Larry, a retired "officer and gentleman," as he dubbed himself in a series of e-mails, a military man who had spent decades in the service. He had first come to this region when he was stationed at Fairchild Air Force Base in nearby Airway Heights. He disclosed also that he had lost a job as an accountant, which occasioned his attending college in his fifties. Much later I learned he was a student of Ayn Rand and a tax resister. He was, in sum, a Libertarian. He had addressed the John Birch Society, whose local chapter rents billboards to seek converts for its opposition to U.S. membership in the United Nations. (Recently elected president George W. Bush was refusing to cough up UN dues; he would later name a UN ambassador, John Bolton, who had published articles opposing the United Nations and would be ousted in short order.)

So I found myself heading a small class, grounded in studies of activism and place, a class that included three retired military people, two of them white males in their fifties, one of them a black female in her twenties. Larry objected vociferously to the assignment for service learning, which I defined as learning that is altruistic. He protested that assigning altruism isn't right, it isn't fair, and it does not constitute sophisticated learning. I tried an example, a tentative assertion that some people turn to teaching as an unselfish endeavor. He scoffed and said teachers get paid too much for teaching for that to be accurately classed as a species of altruism. That was the first day of class; I knew we were in for a wild ride.

Larry's behavior flashed me back to the previous decade, when members of the Aryan Nations had leafleted parked cars at the University of Idaho campus. Racists from outside the campus were aiming for recruits from the student body. At the 1993 meeting of the Rocky Mountain Modern

Language Association in Coeur d'Alene, Idaho, just down the highway from Hayden Lake, where Richard Butler hung his swastikas and lit ceremonial crosses, I had presented a conference paper. The Rocky Mountain Modern Language Association, adopting a theme that year of ethnicity and race, set out to thumb a nose at the racists nearby.

During the keynote address at the conference, four members of the Aryan Nations crashed our meeting, so to speak. They made their way into the ballroom—conspicuously dressed like street punks in leather jackets, high boots, patches, insignias, and berets—and, incidentally, stood behind my folding chair. It was a tense stretch. Finally they did their best to commandeer the discussion and redirect it toward home schooling. One of the four, leather-clad Floyd Cochran, quit the next year when he learned that his comrades would not support his right to keep alive his "defective" cleft-palate child (Hochschild). Something about the purity of the gene pool seemed to be at stake for the Aryans there.

FOR THE TIME BEING

At the same time when I was relinquishing control of my Advanced Exposition class, the Aryan Nations was losing its grasp on the region. (The headquarters at Hayden Lake lay northeast from my office at Eastern Washington University some fifty miles, while the University of Idaho lay southeast some ninety miles.) Here's what happened. A car carrying a mother and son, Victoria and Jason Keenan, was passing the ramshackle compound and backfired. Mistaking the exhaust report for a shot, the guards chased, assaulted, and fired on the family. In the wake of that event, Morris Dees of the Southern Poverty Law Center saw his chance and sued on behalf of the Keenans. He bankrupted the Aryan

Nations. The Reverend Richard Butler had to sell his house and his scattered outbuildings to pay damages. Butler died in 2005. The "world headquarters" of his organization moved to the Midwest.

The territorial imperative that the Aryans hoped to foment in lily-white Northern Idaho proved too slippery to grip, much less to expand to include the western states north of California, including the western Canadian provinces and Alaska. This expansive area was to have been named the Aryan National State. Curiously, it lay along roughly the same lines as Ecotopia, the fictional province based on ecological sustainability and envisioned by Ernest Callenbach in his 1975 novel of that title. Today the West Coast is known as the Left Coast, among some conservatives at least, for its concentration of progressive types.

Something about the far Northwest has attracted radicals for decades. Wobblies, the International Workers of the World, built a short-lived stronghold in Portland, Seattle, Spokane, and Everett in the early decades of the twentieth century. Their heirs today have established alliances with environmental groups, including Roger Featherstone, who in 2001 dubbed himself "kind of a union organizer for grizzly bears" (2). Robert Mathews in the 1980s split off from the Aryan Nations, founded the Order, and "recruited followers to commit murder and robberies here because Butler wouldn't back his call for violence. The cold-blooded killers of Denver radio talk show host Alan Berg were spurred to action here by incendiary speeches delivered by Butler and his comrades in hate" (Oliveria B8). Now the FBI is rounding up participants in the shadowy Earth Liberation Front, whose northwestern adherents have inflicted more than fifty million dollars' worth of economic sabotage on laboratories and housing developments in the American West.

HERE AND NOW

Most of my colleagues adopt an objectivity regarding matters of politics, race, class, gender, economics, and environment. They are aiming for the treasured scholarly disinterest that is believed to characterize good classroom behavior. But I believe that objectivity in such issues is a myth. No crime or liability ought to attach to outspokenness on issues, and I worry sometimes that my colleagues have lost touch with what it means "to profess." At the same time, my students may write whatever they like, no matter how it runs counter to my own philosophies, and I will grade them fairly. Indeed, I will respect them for taking risks, for refuting conventional wisdom and questioning my authority.

Students everywhere should enjoy due process when they think they've been graded or treated unjustly. Students and parents are paying tuition—paying more and more dearly these days—tuition that helps fund the salaries of professors. We ought to be held accountable within education, just as citizens and neighbors ought to be accountable to one another, to other species, and to the sustainability of the water, air, and land.

Environmental studies, the study of place, ought to be also an examination of our bioregions as much for their social and political inner workings as for their topographical and biological parts. If we do not possess a basic understanding of market pressures and population dynamics, of growth management mandates and constitutional arguments about eminent domain (Ring), those of us who teach students about the dynamics of place-based attachments will be giving incomplete views of the complexities involved.

Long ago I lit out for the territories, leaving behind the city of Seattle, where I was born. I chose to work in a conservative region, and that has made all the difference. It has its challenges, this teaching about environment and place to occasionally adversarial audiences, and I confess I sometimes wish that I were once more preaching to the choir in the blue-state half of Washington. But I continue to fight the good fight, partly by asking my students to look their places and circumstances in the face. With caution, skepticism, and a dose of hesitancy, I situate difficult people, ideas, and social movements within the landscape. Custom and culture make them parts of the ecology of the rural West.

COURSE READINGS

Dillard, Annie. *Teaching a Stone to Talk: Expeditions and Encounters.*
Ibsen, Henrik. *An Enemy of the People.*
King, Martin Luther. "Letter from Birmingham Jail."
Thoreau, Henry David. "Walking."
Zinsser, William. *Inventing the Truth: The Art and Craft of Memoir.*

WORKS CITED

Callenbach, Ernest. *Ecotopia: The Notebooks and Reports of William Weston.* 1975. Reprint, New York: Bantam, 1990.
Davis, Claire. *Winter Range.* New York: Picador, 2000.
Diamond, Jared. *Collapse: How Societies Choose to Fail or Succeed.* New York: Viking Penguin, 2005.
Featherstone, Roger. "Earth First! and the IWW: An Interview with Roger Featherstone." *Industrial Worker,* May 1998. http://bari.iww.org/iu120/local/Featherstone/html (accessed March 2, 2001).
Hardin, Garrett. "The Tragedy of the Commons." *Science* 162 (1968): 1243–48.
Lakoff, George. *Don't Think of an Elephant.* White River Junction, Vt.: Chelsea Green, 2005.
Oliveria, D. F. "Inside Hate's Gate." *Spokane Spokesman-Review,* May 27, 2001.
Proulx, Annie. *Close Range: Wyoming Stories.* New York: Scribner's, 1999.
Rand, Ayn. *Introduction to Objectivist Epistemology.* 2nd ed. New York: Penguin, 1990.

Ring, Ray. "Taking Liberties." *High Country News,* July 24, 2006, 8–14.

Schwartz, Delmore. *In Dreams Begin Responsibilities.* Norfolk, Conn.: New Directions, 1938.

"U.S. University Spying Scandal Prompts Resignations." *The Guardian.* January 20, 2006, http://education.guardian.co.uk (accessed January 23, 2006).

Walter, Jess. *Ruby Ridge: The Truth and Tragedy of the Randy Weaver Family.* New York: Regan, 2000.

CHAPTER 14

Weaving Wildness
The Paradox of Teaching About
Wilderness as Place

GREG GORDON

After three days of fruitless searching for wolves, I suggest we sleep in until seven, but the students want to get up again at five-thirty for another try. Although it's mid-July, Yellowstone's Lamar Valley is virtually empty of people, save wolf biologist Rick McIntyre. He informs us that his radio telemetry is picking up a few signals from the Slough Creek Pack. Rick waves at the steep hill behind us and says, "They should be visible from up there."

The university students need no more encouragement; they race up the hill and quickly set up spotting scopes. Clearly visible on the mountainside bench opposite us, four young wolves chase three adult bison, testing them for weakness. Generally, wolves prefer easier prey, such as elk. However, by midsummer the elk have scattered into the high country, and the wolf pups are still too young to travel that far. These sub-adult wolves simply don't know any better than to tackle such formidable prey. The alpha female of the pack watches their antics, with bemusement perhaps. In their romp among the bison, they unveil a weakness in one of the cows; the alpha takes notice and joins the chase.

Possibly the cow had been hit by a car or suffered from some previous injury or malady. Nevertheless, the wolves are on her heels like dogs herding cattle. In less than a minute, the wolves have pulled the bison to the ground and are feasting.

The following dawn we return and so do the wolves. But now a big male grizzly sits chewing on an old elk carcass nearby. The wolves soon eat their fill of bison and head over to torment the bear, rushing in and nipping at him or grabbing pieces of meat from under his nose. The bear lunges at them but won't leave the carcass. Finally, some of the wolves grow bored of this game and flop in the grass. Two others ease in and feed off the carcass side by side with the bear, who apparently decides that chasing wolves isn't worth the energy. Eventually the two wolves rejoin the others and leave the bear to finish his meal in peace. Finally satiated, the grizzly waddles up the hill and into the woods for a nap. As soon as he leaves, the wolves return to the elk carcass.

Meanwhile, four coyotes, a female and three pups, have discovered the bison carcass. Unbeknownst to the coyotes, however, six wolves are feeding on the elk. From our vantage point, the bench opposite resembles a stage where the drama unfolds before us, with the actors unaware of each other. Suddenly the alpha wolf notices the approaching coyotes and begins stalking them. Two of the other wolves follow her lead. The mother coyote raises her head when she spies the wolf. She turns tail just as the wolf bolts for her. The coyotes scatter, the pups going one way and the mom going the other. Despite their superior size and speed, the wolves are laden with full bellies, and the coyotes have a considerable lead. The wolves soon abandon the chase and settle down to guard both carcasses. The coyotes retreat to a nearby knoll and yip at the wolves. The alpha female ignores them but vigorously begins digging up an old coyote den, perhaps just to show them who's boss.

Although Yellowstone's wildlife is world renowned, disappointingly few visitors see the park in its larger context, as the heart of the Greater Yellowstone Ecosystem, home to thousands of humans as well as elk, bison, bears, and wolves. While Yellowstone is firmly lodged in our nation's consciousness, the central question for an educator is how to teach about this as a real place, rather than just another stamp in a national park passport.

After years of teaching university field studies programs, I've realized that to fully know a place we need to develop an experiential understanding of its ecosystem dynamics. Teaching about place requires full physical, emotional, and intellectual immersion, weaving wildness into the fabric of our lives. With wolves now back in the landscape, the ecosystem is one more step toward being restored. But how do we humans fit in?

This is the paradox that these students, engaged in a six-week field studies program through Wildlands Studies (University of California, Santa Barbara), have the opportunity to explore. Through direct observations of rangeland conditions, ungulate migration patterns, and wolf predation, students achieve an experiential understanding of ecosystem dynamics. By participating in wilderness inventories and debates over roadless lands, students also obtain direct knowledge of the complexity of public land issues in the region. Traveling through the wilderness on foot, students gain a physical understanding of the landscape and how animals move through it. Students perceive that vast wilderness is vital to a healthy ecosystem, not just home to wildlife but essential to human nature as well. Yet they also discover that even here, in the wild heart of America, wildlife and wild places are intensively managed for human purposes.

When we return in the evening, more than a hundred people have packed the hillside. Die-hard wolf watchers peer through thousand-dollar spotting scopes, while parents in flip-flops toting small children trek up the hill, hoping to catch a glimpse of the wolves. A family approaches, and I quickly size up the father: red ball cap, well-worn Carharts, cowboy boots, and a faint odor of chewing tobacco and cows. An Eastern Montana ranch hand—or maybe he even has a small herd of his own.

Although the wolves are visible to the unaided eye, I invite his kids to peek through our scope, lowering it to child level. Spellbound, the three kids wait in reverent awe as they take turns watching the wolves play and chew on the bison carcass.

As a local, I've been so inundated with anti-wolf vitriol that I expect the man to launch into a tirade about how wolves are going to sweep through the country and kill all the livestock and game. Instead, he shatters my stereotype as he says from behind his binoculars, riveted on the wolves, "That's about the neatest thing I've ever seen."

As the sun begins to set, the family departs, glowing with pleasure.

What is it about wolves? Why do we stand for hours, enduring weather, boredom, hunger, and lack of sleep just to catch a glimpse of these animals? Sure, they are rare and charismatic, but it seems something happens when we get a close-up view of the lives of animals. When we gaze through a window into the wild, we perceive a glimmer of the world outside of ourselves, independent of human trivialities such as Hollywood and the Dow. Perhaps this is as close to wildness as we can get these days. Yet even here in Yellowstone National Park, the lives of wild animals must conform to human-imposed constraints. Many of the wolves wear radio collars, the bison aren't permitted beyond the park boundary, and bears are removed when managers deem their behavior "inappropriate."

The next day, the students and I hike through the bottomlands of Lamar Valley, assessing cottonwood and aspen regeneration. Recent studies suggest wolves might be responsible for the first cottonwood recruitment in nearly one hundred years. From the 1930s to 1967, the Park Service became so concerned about overgrazing by elk that it actively culled the herds, bringing the population down to 4,000. Yet the deciduous trees were still being browsed so heavily that none reached maturity, and it looked like Yellowstone would soon be bereft of cottonwoods and aspen. In 1967, artificial culling gave way to a new policy of "natural regulation," the idea that wildlife populations would naturally fluctuate and regulate themselves. Indeed, in the years following the cessation of culling, the elk population exploded to 20,000, then leveled off to around 15,000, the point at which density-dependent factors such as forage availability and population density caused elk to produce fewer calves.

To test the impact of herbivores on the range, the park erected several grazing exclosures in 1957. We stop at one such exclosure, a one-acre fenced area. The difference between inside and out appears immediately obvious and profound. Inside stands a thriving aspen grove containing a wide variety of ages. The protected trees send colonizing runners outside the fence, producing numerous aspen shoots. We examine a sapling that stands less than fifteen inches high and age it at fourteen years. For comparison, I indicate another fourteen-year-old tree inside the exclosure that stands taller than eight feet. Clearly, elk herbivory affects aspen regeneration. But what else might be at work here?

"Look at the range from an elk's point of view," I tell the students. "Think of it like a pizza. All spring and summer, there's this nice green grass and the elk are eating the pizza toppings. During the fall the grass begins to die back, and they eat the pizza crust. During the winter, they eat the box."

"So that's when the aspen are eaten?" Angie asks.

"Exactly, and what might that tell you?" I ask, applying the Socratic method, which proves ideal for small-group interactions.

"That there's too much snow?" she responds, fingering the foot-long aspen shoots.

"Well, in some places, but the Northern Range doesn't really get that much snow. What can happen is a freeze-thaw event that creates a layer of ice that the elk can't get through. What might happen then?"

"The elk starve," says Paul.

"Given those conditions, where would you want to spend the winter as an elk?"

"Lower elevations?" says Angie.

"Such as the Paradise Valley?" I prompt.

"Which is full of people," adds Gary.

"Which suggests that this isn't ideal winter range and that the elk would rather be at lower elevation. What does that tell us about aspen regeneration?" I ask, trying to get to the heart of today's lesson.

"That herbivory is a result of the park not being big enough," says Paul tentatively.

"So instead of trying to control elk numbers, the park should have been trying to buy winter range?" continues Angie.

While the thirty-year experiment at artificially reducing elk numbers did little for aspen or cottonwood regeneration, some ecologists have recently suggested that the presence of wolves will succeed where human managers failed. In theory, elk avoid high-risk areas where they might be trapped or lack escape routes. Thus cottonwoods and aspens should escape herbivory in those areas.

To test this theory, we split into five groups to survey the valley's scattered aspen groves. When we reconvene, the students report their findings. Our cursory appraisal finds that all aspen in open areas are heavily browsed, with zero trees able to reach maturity. However, against steep banks and in downed timber, we discover six-foot-high aspen saplings, suggesting that elk indeed spend less time in terrain where they might feel vulnerable to wolf predation.

Over lunch, the students debate the various scientific papers I included as part of the course reading, exploring other factors such as seasonal flooding or climate change that might also explain the ecological state of Yellowstone's Northern Range.

While Yellowstone forms a sense of place on a grand landscape level, just outside the park boundary lies an area popular with local residents for hiking, skiing, hunting, and escaping the crowds. District ranger Ken Britton agrees to show us this corner of the Gallatin National Forest. He holds up the environmental assessment for the Darroch-Eagle timber sale, a two-inch-thick document, and pulls out a map and satellite photo of the proposed sale. Just north of Yellowstone, the sale area also adjoins the Absaroka-Beartooth Wilderness. Ken's photo shows old clear-cuts from the 1970s along the wilderness boundary. This sale would remove most of what is left.

Ken explains a new technique being employed by the Forest Service, leaving islands of undisturbed habitat up to two acres inside the cutting units. He points out the horizontal blue lines marked on trees to indicate the island boundaries on either side of the dirt road where we stand.

"The road runs through the island?" asks Jason incredulously.

"Uh, in this case, yes," replies Ken. "Shall we go see if we can find the island?"

We step off the hot and dusty road, following him into the dense forest, entering a different realm. In just a few meters, the forest has absorbed the heat and noise of the motorized world and envelops us in trees, moss, soil, and living energy. Even our own voices become softer.

We stand next to a large spruce marked with a blue stripe and peer through the forest, trying to find another marker tree. We eventually spot it and walk upslope, then see another marked tree in the opposite direction.

"Is this a corner?" I ask, perplexed.

"Well, this island sort of comes up the drainage and then angles off. It's a bit confusing," Ken admits.

"How is the logger going to know which trees not to cut?" Angie asks pointedly.

"It's confusing," Ken repeats with a shrug.

Finally we give up trying to trace the island and move up to the highest cutting unit, just below the wilderness boundary. Ken tells us that the Forest Service decided to exclude white bark pine from timber harvest since it's a crucial food source for grizzly bears; yet the surrounding lodgepole pines would still be removed. The Forest Service hasn't yet marked the white bark pines, and we scan the treetops for the distinctive five-needled clusters.

"What about all the little trees? Will they just get run over?" Angie asks, indicating the white bark pine saplings amid all the lodgepoles.

"Yeah, we'll just mark the mature trees," Ken replies.

Our final stop is back down the mountain, where logging occurred last summer until a court injunction filed by a local environmental group brought it to a halt. Residents of Gardiner, Montana, have formed a sense of place with this area just as deep and profound as our nation has with Yellowstone. The decisions about logging the area, however, were made in Washington, D.C., despite vigorous local opposition.

Although the students have been in the Yellowstone ecosystem for only a few weeks, they have quickly formed an attachment to this place as well. Their emotions are palpable as we walk up the skidder trail covered in slash, clambering over unwanted logs and ripped-up trees. We reach a bench where all the merchantable timber has been removed. Ken sits on a stump.

"How old do you think that tree was?" Angie asks.

"Oh, about two hundred fifty, three hundred years old," Ken replies.

"It seems absurd to me that the Forest Service is destroying this area, which is in prime grizzly bear habitat, to pay for buying land to protect grizzly habitat somewhere else," Diane says, referring to the land exchange that prompted this sale.

In the late 1990s, Congress negotiated a series of land swaps in the Gallatin National Forest to consolidate public lands and protect wildlife habitat. However, Montana senator Conrad Burns insisted that timber sale proceeds be used to pay for the deal.

Ken responds, "That's what Congress decided to do. The Republicans refuse to release any money from the Land and Water Conservation Fund for outright purchase."

"What's your personal opinion?" asks Angie.

Ken scratches his chin. "Well, I think it's a good sale. I think we could reduce the number of cutting units, especially the higher ones in the white bark pine. But if it doesn't go through that's okay too."

Trained as a silviculturist, Ken believes in a managed forest, yet he also appreciates the value of wilderness. Like most Forest Service employees, he follows the bureaucratic paper trail through this dilemma, ultimately letting the courts make the final decision.

Two days later, we are standing atop Ramshorn Peak (10,296 feet), looking east across Paradise Valley and back toward the Absaroka Mountains. At just under one million acres, the Absaroka-Beartooth Wilderness provides core wildlife habitat for grizzly, wolf, wolverine, and possibly lynx, as well as elk, moose, mountain lion, and mountain goat. Below us, Tom Miner Basin spills into Paradise Valley. We can see how crucial this valley is to wildlife as natural winter range and as a migration corridor between mountain ranges. Unfortunately, as the scattered houses below attest, the valley, a swath of private land bordered by national forest, is in danger of becoming subdivided and developed.

Still draped in snow, the Madison Range dominates the western skyline. I point out the three isolated units of the Lee Metcalf Wilderness, separated by Cabin Creek, a contested roadless area, and the Big Sky ski resort. Vertical stripes on the slopes of Lone Mountain denote ski runs. The road winds back and forth, tracking the sprawling real estate development as it spills out of the valley and creeps upslope. On the other side of the mountain lie the golf course and trophy homes of the private ski resort: the Yellowstone Club, Montana's first gated community, developed by Tim Blixseth, who made his fortune off timber and who profited immensely by the Gallatin land exchange.

Stretching before us to the north is the Gallatin Range, our home for the next two weeks. Our route takes us along the crest of the mountains, enabling us to walk from Yellowstone to the outskirts of Bozeman, Montana. Part of our project will be to document off-road vehicle use in this roadless area in preparation for the upcoming Forest Service Travel Plan, a document that will determine what roads and trails are open to motorized use. The students' comments will become part of the planning process and public record.

On the horizon, we discern the fringes of the Yellowstone ecosystem: the Tetons to the south, the Crazy Mountains to the northeast, the Bridger Range to the north, the Tobacco Root Mountains to the northwest, and the Pioneer Mountains to the west. With the exception of the Tetons, none of these outer ranges contains a single grizzly bear or wolf pack.

Although the view is spectacular, we realize that we are standing in the midst of an ecological island, one that is becoming increasingly fragmented by encroaching development on all sides.

Back at Ramshorn Lake, Alex Phillips of the Montana Wilderness Association hikes in to meet us and discuss the Forest Travel Plan, which would allow motorcycles along the Gallatin Crest and create a snowmobile play area below Ramshorn Peak.

"After your time out here, you guys will know this area better than 90 percent of the people making comments. Even the Forest Service doesn't get out here very much. So what you guys find and comment on will be especially important," she says.

Alex presents background on the Hyalite-Porcupine-Buffalo Horn, a historically complex area. When Northern Pacific was building its railroad across the country, Congress granted the company twenty square miles of land for every mile of track. This resulted in alternating public and private lands, which has been the source of conflict ever since. Indeed, along our hike, we would encounter evidence of this legacy in the form of 640-acre clear-cuts.

Like the residents of Gardiner, the people of Bozeman and Paradise Valley regard the Gallatin Crest as part of their local landscape and care passionately about the place, but they disagree as to how it should be managed. The checkerboard ownership further complicated matters, and Congress punted by granting interim protection as a wilderness study area. While this prohibited logging, the Forest Service believed that motorized use would not impair the area's wilderness character. However, it did not anticipate the explosion in off-road vehicles in recent years.

That evening Alex asks students what wilderness means to them.

"I can't believe how much space there is. You go around a corner and there's more land. It's endless. I never knew there was this much land anywhere," says Angie, who hails from San Diego.

"Wilderness is a place where my mind is at peace and not filled with all the things I have to do. Everything seems so simple out here—eat, sleep, hike. I guess I'm just happier, less concerned about trivial things," says Paul.

"Being out here in all this space with grizzlies around makes me think how nature is so much bigger than us. It's really humbling," adds Gary.

"I feel empowered, like I can make a difference. I thought voting was about all I could do. Now I know I can write letters and that it really does make a difference," says Jason.

"Knowing all the plants makes me feel empowered. I know which berries are edible and their names, and I can share that with others," adds Kirsten.

"When I came here I was really fed up with school, but this is the kind of learning I want to do," says Chris.

"Oh my God, six weeks living in a tent in the woods. I can't believe I made it through this. So many things are not a big deal anymore. Look at my feet—they're filthy, and I don't even care," says Susan, who continuously struggled with the whole outdoor experience.

"I feel more focused and less distracted. I also feel like I've regained that teenage rebellion. I'm angry about things and want to do something about it," says Diane.

Along the Gallatin Crest, we hike through endless fields of alpine wildflowers. They arrive in patches like pointillist paintings, the brilliant yellows of alpine cinquefoil, the uplifting blues of lupines, and the shifting shades of reds of paintbrush. Bistort waves in the breeze like thousands of tiny white Cossack hats.

We experience the landscape through our feet as well as through our senses. Backpacking through the wilderness, we sense how a bear or a wolf might travel, lighting out for new territory as it moves from the core ecosystems into more-human-dominated landscapes. The principles of conservation biology, island biogeography, and habitat fragmentation cease to be abstract when experienced on a physical level.

Despite the commanding view in all directions, Pete declares, "It's weird. This doesn't really feel like a wilderness—I mean, there's the Paradise Valley on one side and Big Sky on the other."

"I imagine that's pretty much how a grizzly bear feels," says Angie.

Between the clear-cuts, the off-road vehicles, and the realization that wildlife habitat is quickly becoming subdivided and fragmented, the students are becoming rather pessimistic, and not without reason.

However, grand schemes are in the works. Applying the concepts of conservation biology to landscape management has provided the genesis for a bold new approach to conservation. The Yellowstone to Yukon Conservation Initiative (Y2Y) would link core ecological reserves via migration corridors that would allow wildlife movement and genetic exchange from Yellowstone to Banff and Jasper National Parks in Canada and on to the Yukon.

At first, the students are a bit skeptical about how this might actually work. After all, we are hiking right through one of the major corridors that would link Yellowstone to the rest of the ecosystem, and we can see clear-cuts on one side and encroaching development on the other. But when I present a few GIS maps of Bozeman Pass (a crucial wildlife corridor) that pinpoint wildlife movements superimposed with land ownership, the students see that not much is required to allow for wildlife migration across human-dominated areas. A combination of conservation easements and wildlife-crossing structures could easily connect the Bridger Range to the Gallatins, which in turn flow right into Yellowstone. Indeed, the U.S. Northern Rockies provide the most optimistic chance for landscape-level conservation.

However, as global warming and exotic species threaten the very icons of American wilderness, like grizzly bears and bison, it becomes clear that wildland protection alone is not enough. Ironically, to preserve wildness we turn to more technology and increased manipulation, whether it be bio bullets for bison inoculation against brucellosis or genetic modification of white bark pine to prevent blister rust. Yet, in this increasingly managed landscape we discover that the wolves turn out to be the best managers of elk after all.

A further irony lies in teaching about wilderness as place. According to the Wilderness Act of 1964, wilderness is supposed to be "a place where man is a visitor who does not remain." While I extol the virtues of behaving as a visitor in wilderness, at the same time I wish to instill a sense of ownership, a psychic attachment to this specific place, this exact ridge if nothing else.

After our discussion, Pete states that nobody really cares about the environment and says how he is being pressured to go into the family real estate development business to build monster homes for people. It would be easy to dismiss Pete, a rich surfer dude from Orange County and a total slacker who has repeatedly said in class discussions that he's just interested in surfing and fishing and nothing else matters. Yet his comments are not entirely without merit, as they do embody many societal norms.

Finally, on the last full day in the backcountry Pete is still the only one who hasn't led a hike. It's his turn, and today's hike is a long twelve miles up and over a two-thousand-foot pass. Given the sudden responsibility, Pete undergoes a personality shift: he comes alive. The previous day he scouted the trail (the only one to have done so). Now he gives people time checks to departure; he constantly consults the maps; he checks in on everyone and leads the group to the divide in half the expected time. That evening everyone says what a great job he did. But what surprised me the most was watching him step around the spiderwebs strung across the trail between stalks of bear grass so as not to disturb the spiders. Everyone behind him followed suit.

COURSE READINGS

Beschta, Robert. *Cottonwoods, Elk, and Wolves in the Lamar Valley of Yellowstone National Park.*

Draffan, George, and Janine Blaeloch. *Commons or Commodity? The Dilemma of Federal Land Exchanges.*

Mattson, David, et al. *Grizzly Bears.*

National Academy of Science. *Ecological Dynamics on Yellowstone's Northern Range.*

Quammen, David. *The Newmark Warning.*

Schullery, Paul. *Bison in Yellowstone: A Historical Overview.*

Smith, Douglas, et al. *Yellowstone After Wolves.*

Terborgh, John, et al. *The Role of Top Carnivores in Regulating Terrestrial Ecosystems.*

Wilkinson, Todd. *Bear Necessities.*

Wilson, E. O. *The Life and Death of Species.*

CHAPTER 15

Teaching About Place in an Era
of Geographical Detachment

HAL CRIMMEL

A powerful spring storm that downed trees along Utah's Wasatch Front is now rolling toward us across the Uinta Basin. Above the rim of Split Mountain Canyon, lightning flickers in the dry air. The students are a little nervous about settling in for the night, so I remind them we're in a rain shadow, where average annual precipitation is only about eight inches, with most falling as snow. But if thunder follows lightning by less than a count of five, I say, leave the tents and get in the van. Neither comment seems particularly reassuring. But so far not a sprinkle has reached us.

Dinosaur National Monument, some 150 miles east of Salt Lake City, is an ecological hub where the Northern Plains, the Rockies, and the Great Basin Desert converge. As such, it would seem to be an ideal choice for studying place. Plants and fish have adapted—often uniquely—to the 211,000-acre monument, which ranges from 4,730 to 9,006 feet in elevation. Rogue ponderosa inhabit canyons thousands of feet below their usual range. The rare Ute ladies' tresses orchid, a federally listed threatened species, survives in Dinosaur. Adaptation continues as more than seventy-five species of invasive plants challenge the distinctive ecology that defines a sense of place here. Tamarisk, for example, a thirsty plague of a shrub, has spread throughout the river corridor, infesting once-open beaches and side canyons and displacing Fremont cottonwoods.

I first became acquainted with Dinosaur National Monument in the early 1990s on a four-day river trip. During the last three years I have frequently returned to research a book about the monument's river canyons. This spring I am teaching creative non-fiction that emphasizes desert place; eleven students are with me for the course's weeklong field component. While we are on location, our goal is to learn about Dinosaur's climate, hydrology, geography, natural history, anthropology, and so forth, and then incorporate this knowledge into place-based creative nonfiction essays.

Once we get away from campus and in the field, the reading—no longer a chore for students—spins itself into gold. I notice a headlamp burning in one tent all night; next morning the student tells me he stayed up until dawn, finishing *The Secret Knowledge of Water,* by Craig Childs. Along with these essays on desert lands, we also read works by Edward Abbey, Ellen Meloy, Terry Tempest Williams, and Ann Zwinger. Though Zwinger's *Run, River, Run* is the only one of the readings actually to discuss Dinosaur National Monument, all teach us about arid and semi-arid lands, and all provide the students with helpful models of place-based creative nonfiction.

Free from cell phones, iPods, the Internet, families, and jobs, the students can focus solely on the high desert surroundings, and even the unfamiliar concepts of natural and cultural history become part of informal discussion. Given the unlikely mix of backgrounds, I'm surprised how quickly the group seems to settle into the routine of camping and how well they get along with each other. Of the eleven, seven are married. Nine are women—four of them are nontraditional students in their thirties and forties. One is pregnant. One has a second-degree black belt. Another, a police record. One a Ph.D. Religious differences, ever a theme in Mormon-dominated Utah, don't seem to be playing a role either. The group shares a rare and genuine camaraderie despite the mix: an atheist, an evangelical Christian, a Catholic, a Unitarian, six Mormons, and a hard-partying Jack Mormon who regales us with Raymond Carver-esque stories about his social life as a tire-store employee.

As for myself (in the interest of full disclosure), place is an obsession, though one I've learned to repress. In graduate school I'd once taught a class on "Sense of Place" perhaps a little too enthusiastically. "If I ever hear the expression 'sense of place' again I will be sick," wrote one woman on her course evaluation. Since then I've tried to resist telling students in the classroom why we should value the idea of place, let alone any particular one, preferring to let places speak for themselves on location.

Yet I've also come to wonder about the role place can play for students living in an era of geographical detachment.

On this trip the students and I are spending six days in the monument. We seem to be guided by biologist E. O. Wilson's observation that people prefer tree-shaded prominences overlooking water. He might have had in mind our cottonwood-lined campground fronting the Green River. For a person captivated by rivers, there are few better places to be in early May, as water is to high desert what sunshine is to Scotland: scarce. It's a long way in any direction to the next significant river: Three hundred miles due north to the Yellowstone. One hundred and fifty miles south to the Colorado. One hundred and fifty miles east to the Cache la Poudre. Four hundred miles due west to the doomed Humboldt.

The Green begins to spike higher in May, primarily from snowmelt carried in the Yampa River, the last remaining major free-flowing tributary of the Colorado. Spring releases from Flaming Gorge Dam near the Wyoming border add to the Green's flow, but the releases, designed to imitate spring floods of the pre-dam era, have done little to achieve the goal of reestablishing beaches, controlling tamarisk, or easing the plight of four species of endangered fish endemic to the Colorado Basin. In the fifty-five miles below the dam before its confluence with the Yampa, the Green's warm-water fishery is near collapse, a victim of introduced northern pike, smallmouth bass, cat-fish, and the unnatural flow pattern and water temperatures created by Flaming Gorge Dam. Distant demand for electricity influences the Green more than anything. "A snake nailed to the earth," Ellen Meloy calls it, a remote river sacrificed to the needs of urban centers in the West (65). This "place," like so many other desert places, is compromised by contact with the outside world, contact that blurs the boundaries between "here" and "there."

Eventually the lightning passes off to the north and the sky clears. Tomorrow will bring sunshine and dry weather: no need to worry about rain anymore. Near the swirling river sleep comes easy.

By next afternoon we're down in a deep canyon along Jones Hole Creek, known for its spring-fed brilliance. It's popular with trout fishermen and hikers. Above the creek at the foot of the canyon wall is Deluge Shelter, a major archaeological site. Excavations revealed evidence of human habitation going back eight thousand years. Today the attraction consists of viewing the bold, reddish petroglyphs and pictographs of animals, humanoids, and geometric designs left by the Fremont, prehistoric peoples who roamed the eastern Great Basin and western edge of the Colorado Plateau over one thousand years ago.

These semi-nomadic indigenous societies can serve as an inspiration for those interested in bioregionalism. The notion that cultures once spent centuries in long intimate and sustainable relationships to place is appealing in our cut-and-run era. Perhaps such relationships existed, though it's hard not to think of Jared Diamond's *Collapse* and the possibility that overpopulation and environmental degradation may have spelled doom for some prehistoric societies.

Even if prehistoric peoples such as the Fremont were economically or spiritually bound to a particular landscape, they still yearned for escape. A few months after the Dinosaur course concluded, I had just finished a talk at the Powell River-Running Museum in the desert town of Green River, Utah, when a retired man stopped to visit. He wanted to discuss his hobby: finding archaeo-astronomy sites, places where the arrangement of rock slabs let a ray of sun mark the equinoxes and solstices. In the country near Green River where Mancos Shale underlies a sparse landscape of greasewood and rabbitbrush, he discovered numerous primitive sundials and petroglyph panels. Most sites, he said, like those across the Colorado Plateau, seem preoccupied with local animals or geography. But other sites, with their fanciful animals, humanoids, and advanced knowledge of astronomy, reflect a desire to escape into the world of imagination and peer into the heavens. Perhaps these, too, are human needs.

The difference between the Fremont era and the present is that technology enables escape from any particular locale, accelerating the process of geographical detachment. In fact, living in place may have more to do with restraint than passion these days. Unprecedented access to distant energy sources, such as natural gas piped across the continent, and to mechanical or electrical technologies means people need not live within the ecological limits imposed by climate and topography. Human hunter-gatherer instincts have not disappeared, but the hunting and gathering is of a different sort. Instead of collecting nuts or following migrating game, consumers gather possessions, degrees, or promotions as though they are seeds.

Thus I wonder if teaching about place is quaint in an era when everyday life often has less to do with the places where Americans live and more to do with goings-on elsewhere. Is maintaining a sense of place or possessing bioregional awareness vestigial as the coccyx, impractical as raising one's own protein? Are these desires simply nostalgic?

Appealing stories abound from an earlier age when families had roots and communities were stable. But let's not discount the power of nostalgia. Many Americans—native and immigrant alike—have a long tradition of displacement and rootlessness. Displacement, whether by force or by choice, is a common denominator among pilgrims, slaves, vaqueros, Indians, pioneers, Chinese laborers, and others. "Place, like memory, grows more potent with distance," writes Tom Vanderbilt, pointing out, for instance, that the last 150 years reveal a history of migration and profound transformation of the American landscape (174).

And most of the students taking the Dinosaur creative nonfiction course are part of that tradition—they've been uprooted frequently.

Those with military ties have moved constantly. Others followed the call of their faith to religious missions abroad. Immigrant parents tied to the old country have hedged others' commitment to place, as seems to be the case with a woman recently returned from Iran. Their classmates are also between cultures and places. One leaves soon for a month of study in Costa Rica; another for four months in Mexico. I am as unfaithful to a particular place as the rest of them. At least thirty different houses, apartments, dorm rooms, and cabins have been my home over the last forty years. I'm preparing for a two-month trip to Austria after the course ends.

Many of the students profess a longing for place, but meaningful, long-term relationships seem impractical, and unlikely in twenty-first-century America. Maybe it's because none of us plan to live near Dinosaur or even in the Uinta Basin. Odds are many of us will eventually relocate from northern Utah. John Daniel's essay "A Word in Favor of Rootlessness" gets at this tension between wanting to root while simultaneously desiring flight. Or, as he describes it, "Marriage to place is something we need to realize in our culture, but not all of us are the marrying kind" (987). And those that are, he warns, can demonstrate an "over-identification with place," where stickers-in-place "run the substantial risk of becoming sticks-in-the-mud" (986).

Knowing that has tempered my enthusiasm for any particular region. I over-identified with place before I came West, shunting people into two categories: those like me who were raised in the sparsely populated country north of the New York State Thruway or in northern New England, and those who were not.

As a graduate student assistant working for the New York State Writers Institute, I once picked up Annie Proulx at the Albany airport.

I knew we had both attended Colby College and had also lived in the remote Maine logging town of Greenville, where I worked as a river guide. Since welcoming visiting writers was part of my assistantship duties, I tried to make conversation, knowing that Proulx, like many of the writers I had met, didn't pull any punches.

"So when did you live in Greenville?" I asked.

"Years ago," she said.

"I lived there, too," I ventured, hoping to bond with another person from the northern forests.

"When?" she asked.

"For three years back in the early 1990s," I said.

"Year-round?" she asked.

"From May until October," I said.

"Summer people," she scoffed.

Summer people! Summer people were members of that polished, well-bred tribe who wouldn't know a power takeoff shaft from a camshaft. They smiled too much and had the odd custom of saying, "Nice to see you," when meeting strangers. They made invitations like, "If you're ever in Greenwich, do call."

Ten years later, I'm not sure I would ever want to go back to so close an identification with any one place. Says John Daniel, "We all need the stranger, the outsider, to shake our perspective and keep us honest" (987). Painful as that can be.

But I can't shake the urge to identify with place. I continue to crave a deep understanding of the places I live and visit, which helps to explain my interest in teaching the Dinosaur course, and it also helps explain how the following misadventure there took shape. One evening our discussion lasted longer than usual, but we still set off on a planned short hike. I hoped we could do the three-mile loop up a dry wash, across badlands, and back down a rocky section before dark. Doing this at a transitional time of day, I hoped, would help students experience an unfamiliar aspect of the monument.

Despite instructions to stay close, Libby, Ryan, and Ben got far in front of the group. I'd hiked the trail before and knew that the section that crossed the badlands—a maze of barren, eroded gullies—was tricky even in daylight. Night was falling fast, and to miss the faint trail in the moonless night would be to vanish into undifferentiated terrain. Telling the rest of the group to wait, I ran ahead to find the three. No sign. I realized we were not going to do this loop before dark, so I sent the others back down the wash toward the van, a mile away. With Kathy, director of university communications and an experienced hiker, I set off to find the missing three. I kicked myself for not playing it safe and doing the hike in late afternoon. Now the dark amplified my worry. Kathy and I found our way through the badlands maze and topped out on a ridge that offered a sweeping view. Nothing. No wink of headlamps, no flashlights bobbing along the trail. Even the group heading back to the van seemed to have been swallowed by the rocks, and unwelcome scenarios began to spin themselves out. What if the three did get lost? What if the other group got lost on their way back? They could find their way in the morning, but the night would be cold and they would have little water. Would I have to call the Park Service and notify the university? I could imagine the wide-eyed FOX News reporter, her voice husky with concern: "We're near the remote town of Jensen, Utah, where tonight a group of Weber State University students have gone missing after an evening hike."

My imagination was transforming this event into a full-blown disaster. As Kathy and I climbed down a steep section, I realized that a slip here and we'd really be in trouble. Savoring the chewing-out the three students would get when I found them provided a mild but satisfying distraction as Kathy and I hunted our way across gulches and sand traps. "The students couldn't possibly have followed this trail," I said. An hour had passed. We had been moving fast, but there was no sign of anyone.

The moment of truth neared as we came back out to the deserted tar road. I thought of the FOX News people, of what was probably bad judgment on my part, prepared for the worst.

"We're sorry," shouted one of the missing three, as we approached the van. Their anxious faces told me there was no need to make a scene. In fact, I wasn't mad—just relieved. They'd been on the trail the entire time; it was the larger group I sent back that should have worried me. They needed only to retrace the dry wash back to the van, but somehow managed to lose their way before getting reoriented.

Though we had camped and hiked for several days, this experience changed perceptions of Dinosaur from a managed park to a wild place, where topography and the cycle of the sun reasserted their dominance. Place, students were surprised to learn, could play a role in the outcome of the day. The students would later write about how Dinosaur's surroundings and the camping, hiking, and rafting helped them challenge phobias and rethink childhood traumas. Some discussed how being alone in nature let them rethink troubles with parents and spouses. Others wrote of reconnecting with the individual selves they had once known before marriage and children.

Nature seems the great healer in a process that tends to refocus our estimation of the magnitude of our problems, and also opens us to the possibilities of place. But what becomes of the connections we've made with Dinosaur when we return to our homes in the urban Wasatch Front? What happens when students return to their jobs in machine shops, tire stores, or at the air force base? What happens when I return to my windowless office and windowless classrooms? I'd like to say we were captivated by place, but our everyday industrialized world leaves little room for that magic to remain.

And sometimes even solitude and scenic grandeur are no guarantee of finding solace in place. For example, when we drove toward the north side of the monument for our hike along Jones Hole Creek, cowboys were moving cattle toward summer grazing allotments on Diamond Mountain. The narrow two-lane blacktop was filled to the horizon with the animals, and soon the van was creeping along through a herd of white-eyed calves and lumbering cows. Like a police escort clearing city streets, a cowboy spurred his horse into the herd, slapping at the mothers with his lariat to open a path for the van.

"You bastard," spat Misty, a twentysomething female student as she flipped off a leathery-faced rancher in this unlikely urban-rural interface.

Later, Ryan Stanger, a student experienced in herding livestock with his working-cowboy father, would write of this incident in his final essay:

Cattle, having thicker skin, can take a much harder blow than these thin-skinned folks in the van. "That swat would have felt like a tap on the shoulder," I assure[d] the group. Wearing a cow-skin jacket and receiving a swat themselves, for the purpose of knowing how to hit a cow, is something all ranchers do before they are handed their whip.

Despite Ryan's reassurance about cows' pain threshold, the chances are slim that Misty embraced the social and cultural aspects of this region. Accepting rural western stock-raising culture might be hard for her, a vegetarian with a strong opinion about what constitutes humane treatment of animals. Adaptation is surely at the heart of accepting place. But it may be impossible to love all places with equal passion.

Still, in this era of geographical detachment, of technologically enhanced semi-nomadism, perhaps our best bet for connecting is to heed the refrain from a tired sixties anthem: "and if you can't be with the one you love, honey, / Love the one you're with."

Or at least try.

That seems to be the philosophy of Ben, the tire-store employee, a born storyteller. One evening he tells the class about skinny-dipping with a date in a pond at a Salt Lake City area dump.

"Ugh," shudders one of the female students, a mother of teenage girls, already horrified by his retelling of amorous behavior in the dump front-end loader's rusty bucket. "How did you know the pond wasn't polluted?" she asks.

"Well, I saw a dog swimming in it a few minutes earlier," he says.

Ben seemed to have found his bliss in the most humble of places. Should this be our goal: embrace all places without reservation?

As the course progresses, the notion of place seems increasingly broad, not exclusively linked to the particular legal entity called Dinosaur National Monument. Three students scramble up a nearby ridge and discover tinajas, indentations in the sandstone that collect rainwater. It doesn't matter that they first learned about this from Craig Childs's *The Secret Knowledge of Water,* in a chapter focused on the Sonoran Desert, an ecosystem profoundly different from Dinosaur's high desert. From *Raven's Exile,* Ellen Meloy's discussion of fish, plants, and animals found in Desolation Canyon, a stretch of the Green River 120 miles downstream, students learn about many of the ecological issues facing Dinosaur. Abbey's sections on the Anasazi help us consider how local Fremont populations, long considered by archaeologists to be "some sort of poor, out-back Anasazi" (Madsen 21), might have interacted with their surroundings.

The essence of place has always evolved as access to game, the availability of water, and types of fauna changed. From the impact of nonnative species to the soaring price of uranium—touching off a new wave of prospecting in the West—that transformation has accelerated in the industrial era. Perhaps in a hundred years, when Salt Lake City is as populous as today's Shanghai and Boise sprawls like New Delhi, we'll come to realize that no amount of concern for preserving particular places can compensate for an ever-burgeoning population. Our refusal to address the P-word means we're facing irrevocable changes in the places where we live.

So what does it mean to live in place in an age of dramatic change and environmental destruction? In this course, the students and I learned about the place known as Dinosaur National Monument. We also learned that a place is less discrete than we might think, as each is overlapped by nearby ecosystems or influenced by distant cities. Place-based teaching, then, might also be understood as an aggregate. Many Americans are suspended between cultures, regions, and ideas. Grounding them in a specific place may be beneficial, but a more effective measure might be to encourage people to see a place—all places—as relevant parts of their lives.

Places and people change. Place-based teaching offers an opportunity to understand how belonging to a region is a process of change, similar to that found in Dinosaur's canyons. Erosion, whether wrought by wind or flash floods, shapes place, as do the shifting fortunes of fish, animals, plants, and trees. Some changes are temporary, some more permanent. Our inheritance has much in common with seeds that are dispersed by wind, water, migratory birds, and animals. Teaching about place can help us to see these temporal connections, connections that can help us rethink what it means to live in place.

COURSE READINGS

Abbey, Edward. *Desert Solitaire.*
Childs, Craig. *The Secret Knowledge of Water.*
Dinosaur National Monument. Map.
Meloy, Ellen. *Raven's Exile.*
Minot, Stephen. *Literary Nonfiction: The Fourth Genre.*
Williams, Terry Tempest. *Refuge: An Unnatural History of Family and Place.*
Zwinger, Ann. *Run, River, Run.*

WORKS CITED

Daniel, John. "A Word in Favor of Rootlessness." In *The Norton Book of Nature Writing*, edited by Robert Finch and John Elder. New York: Norton, 2002.

Madsen, David B. *Exploring the Fremont.* Salt Lake City: University of Utah Press, 1989.

Meloy, Ellen. *Raven's Exile.* Tucson: University of Arizona Press, 1994.

Stanger, Ryan. "Useful, Harmful Water." English 4830: Creative Nonfiction in Dinosaur National Monument. Weber State University, May 2005. Used by permission.

Vanderbilt, Tom. "Is There a 'There' Anywhere?" *Terra Nova: Nature and Culture* 3 (Winter 1998): 167–76.

CONTRIBUTORS

SUEELLEN CAMPBELL, who grew up on what was then the eastern or plains edge of Denver and spent most summers in Colorado's high mountains, now lives about eighty miles farther north in an irrigated valley between the hogbacks and the foothills. She teaches English at Colorado State University, hikes, camps, and travels when she can, reads a lot, and writes books, most recently *Even Mountains Vanish: Searching for Solace in an Age of Extinction*.

LAIRD CHRISTENSEN was born and raised on an ice-age floodplain of the Columbia River and grew up taking for granted the sight of snowcapped volcanoes to the north and east. He still spends summers in Oregon, though he's lived for seven years in western Vermont, where he directs the environmental studies graduate program at Green Mountain College. His poems and essays have appeared in a variety of books and journals, including *Utne*, *Wild Earth*, *Whole Terrain*, and *Northwest Review*.

HAL CRIMMEL grew up on the northern edge of the Adirondacks but now lives along Utah's Wasatch Front. He teaches writing and literature at Weber State University, and served in 2004 as a Fulbright scholar to Austria. He is the editor of *Teaching in the Field: Working with Students* *in the Outdoor Classroom.* His 2007 book, *Dinosaur: Four Seasons on the Green and Yampa Rivers,* is one in a series on desert places.

TERRELL DIXON has lived in bayou country, in Houston on the coastal plain of the Gulf Coast, for more than three decades. He teaches literature and the environment courses at the University of Houston, and he writes frequently about urban nature and its place in American environmental literature. He is the editor of *City Wilds* (University of Georgia Press, 2002).

JOHN ELDER has lived in the Green Mountain village of Bristol since 1973, when he and his wife, Rita, finished graduate school and took up their teaching positions in Vermont. This heavily glaciated landscape, with the drama and beauty of its seasons, has become central to his teaching over the years. In his reading of literature and his writing as well, John has been stimulated by the ways in which human and natural history flow together in this state of towns embraced by unpeopled ridges, stone walls, and cellar holes amid thick forests.

CHERYLL GLOTFELTY, a Montana native, hails from a family of Black Angus cattle ranchers, but she was deported from Big Sky country while still in diapers, destined to spend her formative years in Silicon Valley. She teaches in the literature and environment graduate program at the University of Nevada, Reno. Having coedited *The Ecocriticism Reader* and cofounded the Association for the Study of Literature and Environment, she is now completing a literary anthology of Nevada.

ELLEN GOLDEY grew up in southwestern Ohio's Miami Valley, where the rolling hills are textured with cornfields, cattle, and farm ponds. Her curiosity about the natural world began in youthful exploration of her family's twenty-eight acres of pasture, pond, and woodlot. Now she resides with red clay–tinted shoes in the piedmont of South Carolina, where she is a professor of biology at Wofford College. She has twice been named Wofford's Faculty Member of the Year.

GREG GORDON grew up along Colorado's Front Range. The explosive growth and conversion of open space into starter mansions sent him to Montana's Rocky Mountain Front, where he resides among ponderosa pines, Clark's nutcrackers, and weird rocks. He is currently building a straw bale house powered by solar and wind and trying to learn everything the landscape has to teach. His latest book, *Landscape of Desire: Identity and Nature in Utah's Canyon Country,* further explores our relationship to wilderness.

ROCHELLE JOHNSON grew up just east of Walden Pond and in the forests surrounding Swanzey Lake in New Hampshire, but she now lives north of the Owyhee Mountains in southwest Idaho. She is an associate professor of English and environmental studies at Albertson College of Idaho. Her research takes her to the source of the Susquehanna River and to the Great Meadows of Concord, where she works on Susan Fenimore Cooper and Thoreau.

JOHN LANE's genetic material has been circulating in the Carolina coastal plain and piedmont for more than two hundred years. He now has settled with his family on a piedmont creek, Lawson's Fork, outside Spartanburg, South Carolina. He teaches environmental literature and creative writing there at Wofford College. He writes often, in books and periodicals, about rivers. *Chattooga: Descending Into the Myth of Deliverance River* is now available in paperback from the University of Georgia Press.

PAUL LINDHOLDT grew up on Puget Sound but now lives on the Columbia Plateau. A professor at Eastern Washington University, he has won an Academy of American Poets Award and two awards from the Society of Professional Journalists. He has published some two hundred book chapters, journal articles, essays, columns, reviews, and poems on American studies and American literature, including the recently edited *Holding Common Ground: The Individual and Public Lands in the American West.*

JEFFREY MATHES MCCARTHY grew up on Maine's Casco Bay, and went to college along the Connecticut River at Wesleyan University. Since then it's been mountains and mountains and mountains, with degrees along the way from the University of Edinburgh and the University of Oregon, and a Fulbright Fellowship to the Canadian Rockies. He is chair of environmental studies and associate professor of English at Westminster College in Salt Lake City.

BRADLEY JOHN MONSMA grew up in the farm country, forests, and lakes of northern Michigan and has spent the last seventeen years learning to be at home in the wild and urban landscapes of Southern California. He teaches environmental and multicultural literatures at

California State University, Channel Islands, and is the author of *The Sespe Wild: Southern California's Last Free River* (University of Nevada Press, 2004).

JOHN PRICE grew up in central Iowa and now lives in the Loess Hills of western Iowa, along the Missouri River. He teaches at the University of Nebraska at Omaha and is the author of *Not Just Any Land: A Personal and Literary Journey Into the American Grasslands* (University of Nebraska Press, 2004) and *Man Killed by Pheasant (and other kinships), (Da Capo/ Perseus, forthcoming)*. He has received a fellowship from the National Endowment for the Arts, and his nature essays have recently appeared in *Orion* and *Isotope*.

KENT C. RYDEN divided his growing-up years between the hills of western Connecticut and the glaciated landscapes of eastern Wisconsin, and now he lives a few miles inland from the southern Maine coast. He is the director of the American and New England Studies program at the University of Southern Maine and the author of *Mapping the Invisible Landscape: Folklore, Writing, and the Sense of Place* and *Landscape with Figures: Nature and Culture in New England*.

LISA SLAPPEY grew up in Lithia Springs, Georgia. For years, she studied an obscure geologic formation throughout the Valley and Ridge Province of the Southern Appalachian Mountains. Since 1991 she has lived in Houston, Texas, where she teaches literature and the environment and Native American literature at Rice University. Her essays have appeared in critical collections on John Graves and Ishmael Reed.

ANN ZWINGER lives in Colorado Springs, where in 1970 she wrote *Beyond the Aspen Grove,* the first of her twenty-five books. She was awarded the John Burroughs Medal for Nature Writing for *Run, River, Run* and the Western States Arts Foundation 1995 Award for Creative Nonfiction for *Downcanyon.* She has written for numerous publications, including *Orion, Audubon,* and *Smithsonian.* Edward Abbey called her "the Thoreau of the Rockies." She has taught and given readings across North America.

SUSAN ZWINGER lived in seventeen states but is now settled on an island off the coast of Washington, where she has just completed her fifth book, *The Hanford Reach: The Arid Lands of South Central Washington.* She also wrote *The Last Wild Edge: One Woman's Journey in Search of Ancient Forest.* Her first book, *Stalking the Ice Dragon: An Alaskan Journey,* received the Governor's Author's Award in 1992. She teaches natural history and creative writing workshops in universities, colleges, and institutes across the West.

INDEX